THE CUBAN REVOLUTION

THE CUBAN REVOLUTION

Origins, Course, and Legacy

MARIFELI PÉREZ-STABLE

New York   Oxford
OXFORD UNIVERSITY PRESS
1993

Oxford University Press

Oxford   New York   Toronto
Delhi   Bombay   Calcutta   Madras   Karachi
Kuala Lumpur   Singapore   Hong Kong   Tokyo
Nairobi   Dar es Salaam   Cape Town
Melbourne   Auckland   Madrid

and associated companies in
Berlin   Ibadan

Published by Oxford University Press, Inc.
200 Madison Avenue, New York, New York 10016

Oxford is a registered trademark of Oxford University Press

Library of Congress Cataloging-in-Publication Data
Pérez-Stable, Marifeli, 1949–
The Cuban Revolution : origins, course,
and legacy / Marifeli Pérez-Stable.
p. cm.
Includes bibliographical references and index.
ISBN 0–19–508406–3.
ISBN 0–19–508407–1 (pbk.)
1. Cuba—History—Revolution, 1959.
2. Cuba—History—Revolution, 1959—Causes.
3. Cuba—History—Revolution, 1959—Influence.
4. Cuba—Politics and government—1895–
I. Title. F1788.P455 1993
972.9106′4—dc20   93–19990

2 4 6 8 9 7 5 3 1

Printed in the United States of America
on acid-free paper

In memoriam

Carlos Muñiz (*1953–1979*)
Lourdes Casal (*1938–1981*)
Margarita Lejarza (*1950–1985*)
Ana Mendieta (*1948–1985*)
Mauricio Gastón (*1947–1986*)

Preface

For more than three decades, the Cuban Revolution has challenged minds, engaged imaginations, and aroused passions. From a comfortable Havana suburb in January 1959, I well recall being swayed by the bearded *rebeldes* and the singular euphoria of victory. Although only ten years old, I was very much aware of the extraordinary times that were just beginning. The revolution was a watershed for all Cubans, and I too shared in the pride that the events of 1959 elicited. For my family, however, as for thousands of others, the triumph of 1959 soon lost its aura. We arrived in the United States in 1960 certain that our stay would be temporary.

Growing into adolescence in Pittsburgh during the early 1960s did not nurture my national identity. The civil rights and antiwar movements later in the decade, however, awakened in me a commitment to social justice that led me back to Cuba and the revolution—quite naturally and with joy, but also painfully and with skepticism. I returned to Cuba in 1975 to do research for my dissertation. As I walked the streets of Havana, I strained to remember the city of my childhood. I have visited Cuba frequently since then, and I now have a wealth of memories. Cuba is once again a part of me. This book is more than scholarly endeavor: I have also written it from the heart.

Over nearly 20 years, I have stockpiled debts with family, friends, and colleagues. Some read chapters or the entire book manuscript, others the dissertation—an earlier, remote version. With love, forbearance, and a sense of humor, still others were supportive of me through hard times and encouraging in good times. With most, Cuba and the revolution were constant and vital topics of discussion. In different ways, I thank them all: Mercedes Almánzar, Mercedes Arce, Francisco Aruca, Carollee Bengelsdorf, Adriana Bosch, Lewis A. Coser, Jesús Díaz, María Ignacia Díaz, Nenita Díaz, Jorge I. Domínguez, Mauricio Font, Ambrosio Fornet, Armando García, Mariana Gastón, Robert Greenberg, Hortensia Grey, Rafael Hernández, María Cristina Herrera, Selby Hickey, Susana Lee, Vivian Otero, Louis A. Pérez, Jr., Jorge Pérez-López, Alina Pérez-Stable, Carlos Pérez-Stable, Eliseo C. Pérez-Stable, Eliseo J. Pérez-Stable, Yolanda Prieto, Edwin Reyes, Julia Rodríguez, Albor Ruiz, Helen Safa, Michael Schwartz, Nenita Torres, Miren Uriarte, Juan Valdés, Nelson P. Valdés, María Vázquez, Gloria Young-Sing, Oscar Zanetti, and Andrew Zimbalist. The members of the Proseminar on State Formation and Collective Action at the New School for Social Research coordinated by Charles Tilly and Richard Bensel provided me with incisive criticism on a draft of Chapter 2. Memories of my grandmother, Beba Moya de Díaz, comforted me throughout the writing of this book. I am especially indebted to Jay Kaplan for the happiness and intellectual companionship he has brought into my life.

I am likewise grateful for the professional support I have received from various quarters. The administration of the State University of New York, College at Old Westbury and my colleagues in the Program in Politics, Economics, and Society manifested uncommon understanding of the time it took me to finish my dissertation. Sarah M. Tenen and Laura Healey, PES Program secretaries, were always responsive to my many requests. The Department of Sociology of the State University of New York at Stony Brook afforded me a congenial intellectual atmosphere to finish my doctorate. At the Old Westbury library, Noreen Ashby and Debra Randorf obtained materials for me through interlibrary loan that proved invaluable in completing the book. I am equally appreciative of the patient help I received from the staffs at various research libraries: the New York Public Library, the Hispanic Division at the Library of Congress, the University of Miami, and the Biblioteca Nacional José Martí. I am thankful as well to the Foreign Relations Ministry, the Central Organization of Cuban Trade Unions, and the Federation of Cuban Women for sponsoring my trips to Cuba and facilitating my research in its various stages. Without Nota Bene and the helpful assistance from the Dragonfly Software staff, the manuscript might at times have been lost. Michael Taves also provided me with computer guidance at crucial moments. At Oxford University Press, Nancy Lane, Edward Harcourt, and Ruth Sandweiss ably assisted me in the final preparation of the manuscript.

The Ford Foundation supported my research trips to Cuba in 1975 and 1977. In 1987, I obtained a Graduate and Research Initiative Award from the State University of New York to work in the Library of Congress. I received a 1987 Recent Recipients of the Ph.D. Fellowship from the American Council of Learned Societies, funded in part by the National Endowment for the Humanities. In combination with a 1988–1989 sabbatical leave from Old Westbury, the ACLS fellowship allowed me the time to revise the dissertation and begin the book. I gratefully acknowledge all financial support. Parts of earlier versions of Chapters 3 and 7, and of Chapters 6 and 7 were, respectively, published as "Charismatic Authority, Vanguard Party Politics, and Popular Mobilizations: Revolution and Socialism in Cuba," in *Cuban Studies/Estudios Cubanos* 22 (1992), and " 'We Are the Only Ones and There is No Alternative': Vanguard Party Politics in Cuba, 1975–1991," in Enrique A. Baloyra and James A. Morris, eds., *Contradiction and Change in Cuba* (Albuquerque: University of New Mexico Press, 1994).

The Cuban Revolution: Origins, Course, and Legacy is dedicated to five friends. In April 1979, Carlos Muñiz was assassinated in Puerto Rico at the behest of Cuban-American terrorists bent upon undercutting the rapprochement between the Cuban government and the Cuban-American community. Carlos was president of Viajes Varadero, a travel agency specializing in family reunification visits to Cuba. After a long illness, Lourdes Casal died in Havana in January 1981. She was a wiser and loving older sister to me: I will always miss her vulnerable humanity, her *joie de vivre,* and her monumental intelligence. In January 1985, Margarita Lejarza passed away in Miami. I cherish the memory of her smile and her voice as she sang and played the guitar. Under incomprehensible circumstances, Ana Mendieta died in September 1985. She was strong-willed, fiercely proud of her *cubanía,* and well on the way to leaving her mark as an artist. Mauricio Gastón died in September 1986. He too was *un hombre sincero.* The Puerto Rican community in Boston and the Cuban Revolution motivated his commitment to social justice and engaged his considerable intelligence.

With Carlos, Lourdes, Margarita, Ana, Mauricio, and dozens of other progressive Cuban-Americans, I endeavored during the 1970s and 1980s to support the Cuban Revolution and to ease the estrangement between Cubans in Cuba and in the United States. I then looked forward to the day when sanity and respect assumed their long overdue dimension in relations between the Cuban and U.S. governments and between Cubans abroad and on the island. I still do, but now the terms of that rapprochement may well be very different from what I had earlier anticipated. The violent downfall of the Cuban government and the complete disavowal of the revolutionary legacy are today distinct possibilities. Moderation and compromise in Havana, Miami, and Washington may yet preempt that outcome. I hope so.

New York   M.P.S.
March 1993

Contents

Tables

Acronyms

AJR	Association of Rebel Youths
ANAP	National Association of Small Peasants
ANIC	National Association of Cuban Industrialists
CC	Central Committee
CDR	Committees for the Defense of the Revolution
CMEA	Council for Mutual Economic Assistance
CNOC	National Confederation of Cuban Workers
CTC	Central Organization of Cuban Trade Unions
DEU	Directorate of University Students
DRE	Revolutionary Student Directorate
EAP	Economically Active Population
FEU	Federation of University Students
FMC	Federation of Cuban Women
FNTA	National Federation of Sugar Workers
FOH	Front of Humanist Workers
GNP	Gross National Product
GSP	Gross Social Product
INRA	National Institute for Agrarian Reform

JUCEI	Local Commissions for Coordination, Implementation, and Inspection
JUCEPLAN	Central Planning Board
OPP	Organs of Popular Power
ORI	Integrated Revolutionary Organizations
PCC	Cuban Communist Party
PRC	Cuban Revolutionary Party
PSP	Popular Socialist Party
PURS	United Party of the Socialist Revolution
SDPE	Economic Management and Planning System
UJC	Communist Youth Union

THE CUBAN REVOLUTION

Introduction

"This time the revolution is for real!" Fidel Castro declared upon entering Santiago de Cuba on January 1, 1959. Few Cubans then pondered what a real revolution was and what its consequences would be. Almost all were elated with the downfall of Fulgencio Batista. Cubans from all walks of life exuberantly embraced the young Fidel and the *rebeldes*—the bearded guerrillas who had led the insurrection against dictatorship and now embodied renewed hope and rekindled pride. Two years later none would doubt the revolution was, indeed, for real. The new government had undertaken a radical transformation of Cuban society. On April 16, 1961, Fidel Castro proclaimed the socialist character of the revolution. A day later, a force of U.S.-supported Cuban exiles landed at Playa Girón (Bay of Pigs). Within seventy-two hours, the invaders were routed. The revolution was not only real; it would also survive.

The origins and development of social revolution in Cuba are the subject of this study. In the late nineteenth-century struggles against Spain, many Cubans forged a commitment to national independence and social justice that served as the basis of their radical nationalism. Between 1902 and 1958, the Cuban Republic disappointed the *independentista* mandate: sugar monoculture and dependence on the United States compromised it. Radical nationalism, however, did not dissipate. Intermittently and in various ways, social forces and political groups appealing to its tradition challenged the foundations of the republic and eventually succeeded in establishing themselves as a credible alternative. Thus, *The Cuban Revolution: Origins, Course, and Legacy* subscribes to the main proposition of contemporary Cuban historiography: the origins of the revolution lie in the independence movement against Spain and the frustration of its aspirations in the Cuban Republic.[1]

A radical interpretation of Cuban history is quite persuasive. Alternative interpretations have yet to be articulated with comparable coherence and suggestiveness.[2] Teleology, nonetheless, weakens its analysis: it portrays the Cuban Revolution as the inevitable conclusion of a hallowed destiny. Moreover, because of the revolution, the past acquires the logic of radical nationalism and, consequently, contingencies are eliminated. The past, however, was not mere prelude to the revolution; it harbored alternatives that were never fulfilled. Cuban society allowed the revolution to happen but did so after other paths were not taken. This study is written with attention to emergent logic and frustrated contingencies and, thus, seeks to modify the linearity of Cuban historiography.

Nonetheless, the revolution highlights the forces of radicalism and nationalism in the Cuban past. In 1868, the patrician Carlos Manuel de Céspedes freed his slaves and took up arms for *Cuba libre* (free Cuba) against Spain. Other landowners and slaveholders followed his lead. Though poor whites and free blacks also joined the separatist effort, creole propertied interests predominated in the leadership of the movement. Ten years later Spain prevailed over the insurgents and peace returned to Cuba. During the 1890s, José Martí resumed the cause of independence from the United States. In 1892, he founded the Cuban Revolutionary Party (PRC) with the support of prosperous creoles and working-class Cubans. A precursor of twentieth-century national liberation movements, the PRC promoted unity among all Cubans in order to achieve independence. Once the republic was established, however, liberal visions of *patria* (homeland) coexisted uncomfortably with radical aspirations of social transformation.[3]

In 1895, the PRC launched the second war of independence, and shortly thereafter Martí died in battle. Within two years, the *Ejército Libertador* (Liberation Army) had nearly defeated the Spanish forces: the insurgents controlled the countryside and were preparing an assault on the major cities. Dismayed by the prospect of a republic installed under the aegis of the Liberation Army, some creole property owners encouraged the United States to intervene. Their interests coincided with those of nascent U.S. imperialism; Martí and the popular sectors in the PRC had adamantly opposed U.S. intervention, fearing Cuban independence would be compromised. In 1898, the United States entered the war against Spain and began a four-year occupation of Cuba. In 1899, the *Ejército Libertador* was disbanded. On May 20, 1902, the republic was proclaimed under a new constitution and the Platt Amendment, which allowed the United States to intervene in Cuban affairs whenever order was threatened. Indeed, Cuban independence was compromised.

The republic also thwarted the social justice aspirations of the independence movement. Foreign capital, largely from the United States, owned most of the national wealth and primarily benefited from the rapid sugar expansion of the early twentieth century. Only when profitability slackened after the 1920s did Cubans gradually acquire a majority ownership of the sugar industry. Sugar production, however, stagnated: the population had doubled, but sugar output during the 1950s barely surpassed that of the 1920s. Moreover, sugar still accounted for the bulk of export earnings, and the United States continued to represent its most important, if declining, market. The centrality of sugar reinforced a vicious circle: without sugar, there was no Cuba, and there was no sugar without the U.S. market.

Trade reciprocity—preferential tariff treatment for Cuban sugar in the United States in exchange for reduced customs duties on U.S. exports to Cuba—further favored sugar at the expense of other products. Sugar interests resisted efforts to protect the domestic market for national industry, fearing U.S. retaliation against them. Economic diversification faced powerful obstacles. The sugar industry, moreover, used seasonal labor and controlled the most fertile land, resulting in high unemployment and land concentration. Sugar preeminence reinforced dependence on the United States, curtailed economic growth, and restrained standards of living.

Thus, national sovereignty and the struggle for social justice were the two pillars of radical nationalism. The nineteenth century forged its tenor, the twentieth its intransigence. The republic frustrated its aspirations, bolstered its contentions, and enhanced its credibility. By the 1950s, the nineteenth-century cry of *independencia o muerte* (independence or death) had become *libertad o muerte* (liberty or death). After 1959, *patria o muerte* (homeland or death) would express the nearly one hundred years of struggle for national sovereignty. Socialism would become the conduit to realize social justice. That radical nationalism retained relevance in the republic and emerged as a viable alternative in 1959 was due in no small measure to the complexion of Cuban society and the crisis of political authority.

The year 1959 is the "great divide" in the study of Cuba. Numerous scholars have variously analyzed the social revolution and Cuban socialism.[4] The nature of Cuban society before 1959 is, however, a rather neglected theme in the literature. Caricatured views of prerevolutionary Cuba have too often sustained explanations for the origins of the revolution.

During the 1950s, Cuba ranked among the top five countries in Latin America on a wide range of socioeconomic indicators such as urbanization, literacy, per capita income, infant mortality, and life expectancy. High levels of modernization, the active participation of the middle sectors against the dictatorship of Fulgencio Batista, and the class origins of the opposition leadership inspired the characterization of society and the revolution as middle class. Political reform—the restoration of constitutional government and the curtailment of corruption in public administration—was necessary to realize the "take-off" of the Cuban economy. Because the suppression of capitalism was never the stated objective of the opposition movement, these interpretations attribute the subsequent turn toward socialism to the machinations of Fidel Castro.[5]

A contrasting view emphasizes stagnation as the principal feature of Cuban society before 1959.[6] After the 1920s, sugar monoculture and the concomitant dependence on the United States made it impossible to sustain growth and promote diversification. Cuban modernity masked profound inequalities between urban and rural areas and coexisted with high levels of unemployment and underemployment. Widespread corruption and indecisive leadership, moreover, mired the political system. Socialism was seen as inevitable in order to break the stranglehold of economic stagnation, social backwardness, and political corruption. According to this view, the character of Cuban society—not the leadership of Fidel Castro—was the decisive factor in understanding socialist transformation after 1959.

The social forces interacting in the revolution and supporting the subsequent radicalization were, however, a matter of debate. For some scholars, peasants were the key to the revolutionary victory.[7] For others, radicalized sectors of the middle class mobilized the peasantry to bring the revolution to power.[8] Some saw a coalition of the marginal and rootless across classes as the spearhead of radical change.[9] Still others argued that the Cuban Revolution was "declassé" and thus conducive to a "left-Bonapartist" mediation among classes.[10] Orthodox Marxists contended that an alliance of workers and peasants sustained socialist transformation.[11] One author asserted that the working class, albeit relatively insignificant in the anti-Batista movement, was decisive in consolidating the revolution.[12] Implicit in these various propositions was the question of the kind of society Cuba was before 1959.

That the literature proposes all of these answers and also argues for the middle-class origins of the revolution is an indication of how imprecise the understanding of Cuba before 1959 is. Moreover, the research for most of these analyses was conducted more than two decades ago. Never fully investigated, the nature of the old Cuba seems to have been long forgotten. Analyzing the outcomes of revolution has overwhelmed the field of Cuban studies. The origins of revolution have been sidelined or dismissed with broad generalizations. In 1974, Sidney Mintz noted that the various roles attributed to peasants and rural workers in the revolution required a clearer structural conceptualization of rural Cuba.[13] His call has largely gone unheeded. Not just rural Cuba but prerevolutionary Cuban society is uncharted terrain for sociological analysis. Overcoming the dichotomous characterizations of the old Cuba as either on the verge of "take-off" or irreparably stagnant is imperative. The "modern" profile of Cuba before 1959 needs to be integrated into a structural analysis of the origins of the revolution.

Unlike other major social transformations, the Cuban Revolution has not elicited systematic historical inquiry.[14] This study aims to bridge the "great divide": the social revolution of 1959 is analyzed in view of the "emerging" crises in Cuban society over the six decades after the inauguration of the republic in 1902.[15] Social revolution and the ensuing radical transformation of Cuban society were neither inevitable nor aberrational. The old Cuba sheltered these options as well as others that were never or only partially realized. Sociopolitical dynamics, however, explain how Cuban development eventually followed the paths of revolution and socialism rather than some variant of dependent capitalism. Cuban society provided the propitious context in which the revolution was made and socialism became a viable option. Indeed, even if not exactly at their discretion, people do make their own history.[16]

During the twentieth century, six factors interacted to render Cuba susceptible to radical revolution: mediated sovereignty, sugar-centered development, uneven modernization, the crisis of political authority, the weakness of the *clases económicas* (economic classes), and the relative strength of the *clases populares* (popular sectors).[17] In crucial ways, sugar production established the fundamental logic of Cuban politics. *Sin azúcar, no hay país* (without sugar, there is no nation)—as José Manuel Casanova of the Sugar Mill Owners Association frequently affirmed— succinctly summarized it. The economy revolved around sugar and the reciprocity agreements between Cuba and the United States that reinforced monoculture and prevented diversification. Nonetheless, the capital-intensive sugar industry with its use of wage labor and the intimate ties it engendered with the United States also sustained the relative, if uneven, modernization of Cuba.

Sugar monoculture constituted the structural context that allowed social revolution to happen. Sociopolitical dynamics explain how it was actually made. Until 1934, when Franklin Delano Roosevelt abrogated the Platt Amendment, political elites strove to satisfy the United States just as often—in some instances, more often—as they did domestic constituencies. After 1934, sugar, with its dependence on the U.S. market, continued to define national parameters and limit development options. During the 1930s, social upheavals engulfed Cuban society and brought the political order of the early republic to an end. Their outcome was neither social revolution nor government reaction. The Constitution of 1940—not unlike the constitutions of Mexico (1917) and Bolivia (1938)—embodied a social compromise protecting private property, sanctioning an interventionist state, endorsing agrarian reform, and promoting a host of social rights. The 1940s, however, failed to consolidate that compromise. In 1952, Fulgencio Batista overthrew Carlos Prío, ending twelve years of constitutional government. Compounded over six decades, a crisis of political authority marked the Cuban state. By the end of the 1950s, this crisis was a crucial factor in explaining the relative ease with which the anti-Batista movement gained power.

The Cuban Revolution highlights the importance of social classes in the breakdown of the old Cuba and the making of the revolution. It emphasizes the role of the *clases populares,* especially the unionized working class, in the mounting crisis of Cuban society. It similarly underscores the inability of the *clases económicas,* especially the nonsugar sectors, to implement a development program to alleviate unemployment and redress the pervasive sense of insufficiency resulting from uneven modernization. The dynamics of state power underscored the political crisis. Between 1902 and 1958, the state responded to the imperatives of sugar. Until the 1950s, however, Cubans owned a minority of the sugar mills.

After the 1930s, the Central Organization of Cuban Trade Unions (CTC) constituted a powerful interest group whose demands the state satisfied up to a point—the price exacted for social peace—resulting in a relatively high cost of labor. Indeed, Cuban capitalists often faced rather unfavorable conditions for the conduct of business. Organized in the National Association of Cuban Industrialists (ANIC), the nonsugar sectors were the most neglected by state policies: their interests were nearly always subordinated to those of sugar. Moreover, the CTC was a more influential organization with a larger constituency than ANIC and, consequently, the state frequently favored labor in conflicts with capital.

By the 1950s, classic dependence on sugar was exhausted. Monoculture could no longer sustain economic growth, as it had until the 1920s and during wartime. This study argues that the transformation of Cuban capitalism might have led to a missing model of Latin American development. Obviously, this argument cannot be proven, since it is about a history that never happened. Its usefulness lies in its power to suggest and provoke. Caricatures—like those of "take-off" or stagnation—are, however, more misleading. The kind of society that Cuba was is a crucial question in the field of Cuban studies, and Cubanists need to answer it satisfactorily. This study argues that Cuba was on the verge of a transition from classic dependence to a new form of dependent capitalism. The interactions of the state, foreign capital, the dominant classes, and sectors of unionized labor were pointing to what I am calling "tropical dependent development." The alliances that might have sustained the transition, however, never coalesced.

During the 1950s, the politics of the old Cuba was likewise practically spent. Twenty years after the social upheavals of the 1930s and the abrogation of the Platt Amendment, the Cuban political system had yet to settle. The three administrations of the 1940s miscarried the consolidation of constitutional rule. Although elections regularly and with relative honesty determined the political leadership, corruption and malfeasance marred the process of governance. Political parties were generally weak, responding more to personalities than platforms. Violence too often settled political scores as *grupos de acción* (action groups) increasingly ruled the streets of Havana and other cities. When Batista overthrew Prío in 1952, few mourned the passing of Prío's government. The Constitution of 1940, however, became the symbol of the highest expectations of the citizenry, and its restoration soon developed into the rallying cry of the opposition movement.

The character of the anti-Batista forces is a central element in understanding the breakdown of the old Cuba and the making of the social revolution. This study contends that the opposition movement cannot be divorced from the underlying societal dynamics, that is, from the emerging crises over the course of the twentieth century. The pledge to restore the constitution needs to be seen in view of the significance of 1940 in relationship to the revolutionary upheavals of the 1930s and the actual experience of representative democracy. The magna carta symbolized social justice as well as formal democracy. Moreover, Batista's resistance to calling elections undermined the moderate opposition and bolstered the July 26th Movement led by Fidel Castro. Because armed insurrection rather than negotiations ended the dictatorship, institutional processes were further debilitated. Even though it mobilized a broad spectrum of Cuban society, the leadership of the anti-Batista movement had, at best, tenuous bonds to the old Cuba. By January 1, 1959, when the *rebeldes* came to power, the forces that might have contained the onslaught of revolution had been previously weakened. Moreover, Fidel Castro was not predisposed to compromise the "real" revolution in order to appease the *clases económicas* and the United States.

Radical nationalism guided Fidel Castro and the *rebeldes* throughout the crucial year of 1959, when the social revolution was actually made. They enjoyed overwhelming popular support. Cuba had never had a leader so beloved: *el pueblo cubano* (the Cuban people) truly believed in him and his visions. Castro exercised his extraordinary leadership to marshal the popular ground swell and consolidate a radical revolution. Throughout 1959, three factors interacted to foster a process of radicalization. From the outset, a dynamic developed that allowed the revolutionary government to mobilize the *clases populares* and reinforce their newly acquired sense of empowerment. Demands from below for higher wages, employment, and other benefits were often satisfied. Agrarian and urban reforms symbolized the willingness of the government to fulfill popular claims. The *clases económicas* and the United States became increasingly alarmed. Fidel Castro emerged as the bulwark of the revolution—the mediator between the people and the government. The first signs of U.S.

hostility, moreover, fueled radical nationalism and bolstered the centrality of Fidel Castro. During 1959–1960, popular empowerment, charismatic authority, and U.S. aggression crystallized revolutionary politics.
The dynamic of Fidel-*patria*-revolution was a consequence of the processes of 1959 and has since remained at the heart of politics in Cuba.

   The Cuban Revolution maintains that the impetus to radicalize the
revolution came from the interaction of social forces in the course of
1959, the willingness of the revolutionary government to respond in
favor of the *clases populares,* and the inability of the *clases económicas* to
mobilize a counterresponse. The process of radicalization also implied
some form of confrontation with the United States. The revolutionary
leadership chose to encourage popular pressures from below; make few
concessions to the industrialists who were initially considered allies;
resist all compromises with the dominant sectors of the *clases económicas;*
and defy the United States. They could have restrained popular demands, appeased the *clases económicas,* and mollified the United States.
During the 1950s, the National Revolutionary Movement had followed
such a course in Bolivia. Had the Cuban leadership moderated its radicalism, the revolution would have had a different outcome. It did not
and, consequently, the imperative became survival in the face of opposition from the United States, the *clases económicas,* sectors of the middle
class, and even the more privileged among the *clases populares.*

   The Cuban leadership focused on three crucial elements in consolidating its rule: developing the economy, seeking new international
allies, and constituting a new political authority. Redressing past inequalities and establishing social justice as the guide of economic development have been central concerns of Cuban leaders. When economic
growth and social justice have been at odds, they—especially Fidel
Castro—have preferred the latter. Forging a new *conciencia* (consciousness) based on collective well-being has been essential to their economic
development strategies. Moreover, they have considered that defending
the nation against the United States requires a steely national unity,
possible only if social justice, that is, equality among Cubans, is promoted.

   Affirming national sovereignty against the United States, however,
was possible only because of the support of Soviet Union. New ties of
dependence enabled the Cuban Revolution to survive the U.S. embargo,
achieve impressive gains in social welfare, and attain modest, if erratic,
rates of economic growth. Moreover, Soviet mentorship buttressed national security by supplying free armaments and training military personnel. Last, the Soviet Union offered Cuban leaders models of socialism
and one-party politics that in the early 1960s appeared to be feasible
alternatives to capitalism and representative democracy.

   In 1959, the Cuban leadership rapidly established new patterns of
governance: Fidel-*patria*-revolution emerged as the logic of revolution-

ary politics. *The Cuban Revolution* examines the constitution of a new political authority and its subsequent exercise in the relationship of the Cuban Communist Party (PCC) to the trade unions and the Federation of Cuban Women (FMC). It probes the tensions between institutional politics and the underlying dynamic identifying the nation, the leadership of Fidel Castro, and the revolution as inseparable. This study, moreover, argues that, sociologically speaking, Cuba is no longer in revolution because the social transformations that changed the basis of political power occurred during the 1960s.

In the 1990s, the end of the cold war, the downfall of the Eastern European Communist parties, and the dissolution of the Soviet Union have resulted in a decidedly inauspicious international environment for the Cuban government. Moreover, the United States is tightening the economic embargo in the hope of finally attaining the Cuban government's overthrow. Nonetheless, the current situation also needs to be understood in terms of national developments. Domestic and international factors have combined to undermine the long-term prospects of socialism and the Cuban government.

Cuba is a Latin American country. That incontrovertible fact, however, is not so evident in the field of Latin American studies. The revolution has tended to isolate Cuba from the literature on Latin American development. The reasons are understandable. When the Cuban Revolution came to power, the fields of development and Latin American studies were markedly different from what they are today. Thirty years ago, modernization theories dictated notions of development, and Latin American studies were an incipient field. The 1960s—shaped, in part, by the experience of the Cuban Revolution—transformed the study of development and launched interdisciplinary approaches to Latin America.

Dependency and world-system theories successfully challenged the modernization paradigm that had provided accurate descriptions of development processes but had failed to explain them analytically.[18] The new theories focused on the weight of external factors and the links between core and periphery in the world economy to explain underdevelopment. Many *dependentistas*—initially of Latin American origins or specialists on Latin America—advocated socialist transformation to overcome underdevelopment.[19] Because Cuba represented their prescriptions, *dependentistas* spent little analytical attention on the origins and outcome of the revolution. More recently, a "new comparative political economy" has combined historical perspective, the comparative method, and the interactions between states and markets to study development processes.[20] But Cuba is almost nowhere to be found in the postmodernization literature.[21]

Over the past 30 years, Latin American studies have significantly contributed to broadening the field of development. The research of sociologists and other social scientists specializing in Latin America has been definitive in overcoming some of the shortcomings of dependency and world-systems analyses. The study of Cuba has again remained largely outside their purview. Cardoso and Faletto, for example, barely mention Cuba in their seminal work on Latin American development.[22] Portes and Walton do not address processes of social change in Cuba in their analysis of labor, class, and the international system.[23] De Janvry does not discuss Cuba in his study of agrarian reform in Latin America.[24] Cuba was a "natural," but was not included in Bergquist's book on labor and the export sectors.[25] Similarly, Berins Collier and Collier do not consider the Cuban experience in their comparative analysis of state policies toward the labor movement and the consequences of those policies for the political systems of eight countries.[26] Di Tella refers to Cuba only in passing in his theoretical discussion of Latin American politics.[27] Sheahan's book on patterns of Latin American development is the exception that proves the rule: it compares Cuba since 1959 and Peru under Velasco Alvarado in terms of economic strategies, repression, and social justice.[28] There are no works on Cuba comparable to those of Evans on dependent development in Brazil, Becker on the Peruvian bourgeoisie, Gereffi on the pharmaceutical industry in Mexico, Zeitlin on Chilean capitalists, Font on the Brazilian coffee sector, and Gallo on fiscal policy and political stability in Bolivia, to name but a few.[29]

Thus, the fields of Cuban studies and Latin American development have mostly evolved parallel to each other. Most Cubanists have implicitly adhered to the premises of modernization theory. In general, ahistorical perspectives and unabashed empiricism have marked Cuban studies. Specialists on Cuba almost seem to be saying the facts speak for themselves. Explanations of change have too often been limited to a given conjuncture such as the failure of the 1970 sugar harvest and the current Cuban predicament in the post–cold war world, or an emerging crisis in international relations such as the Missile Crisis of 1962, Cuban support for the government of the Popular Movement for the Liberation of Angola during the 1970s, and Havana's assistance to Central American revolutionary movements during the 1980s. Elites are usually seen as the sole political actors; the actions of ordinary Cubans are rarely identified. The links between society and politics are not prominently explored.[30]

The Cuban Revolution hopes to open new vistas in the study of Cuba. I have attempted to integrate past and present through the prism of postmodernization theories. I have likewise endeavored to bring as much evidence as I had at the time of writing to sustain my arguments. Much, I know, remains to be done: a historical sociology of Cuba is still incipient. Nonetheless, I trust I have raised some of the right questions. I also hope this study entices other social scientists, especially Latin Americanists, to take the Cuban experience more fully into account in their exploration of social change. The chapters that follow are almost exclusively about Cuba. Although the narrative is theoretically and analytically informed, I have not engaged the literatures on revolutions and Latin American development, nor have I made systematic comparisons other than the evident one over time in Cuba. Even so, I believe this study makes a modest contribution to both literatures.

Chapter 1 overviews the patterns of Cuban development between 1902 and 1958, with special emphasis on sugar, relations with the United States, the emergence of reformism, and the role of the state. Chapter 2 analyzes politics in Cuba before the revolution. The dynamics of the Plattist republic, the social upheavals of the 1930s, the compromise of 1940, the years of constitutional government, the Batista dictatorship, and the formation of the opposition movement are analyzed. Chapter 3 focuses on the years 1959–1961 and the radicalization of the revolution; it is the heart of the book. Chapter 4 presents a review of development strategies and socioeconomic performance since 1959. Chapter 5 deals with the formation of the Cuban Communist Party and its relations with the CTC and the FMC in view of the revolutionary leadership's efforts to constitute new forms of political authority. Chapter 6 continues the analysis of the PCC and the two mass organizations during the period of institutionalization (1970–1986). Chapter 7 reviews the process of rectification and the crisis of Cuban socialism in the post–cold war world. The conclusion summarizes the main themes and offers some perspectives on the future of Cuban development.

CHAPTER 1

Mediated Sovereignty, Monoculture, and Development

> Without sugar, there is no nation.
>
> *José Manuel Casanova*
> *Sugar Mill Owners' Association*
> *1940s*

> Because of sugar, there is no nation.
>
> *Raúl Cepero Bonilla*
> *Cuban economist*
> *1940s*

Cuba before the revolution was rather unique in Latin America. Sugar monoculture appeared to cast Cuban society in the mold of a foreign-dominated, enclave economy. Until the 1950s, U.S. capital held majority interest in the sugar industry. Cuba-U.S. trade reciprocity, moreover, undermined the diversification of the Cuban economy. Appearances, however, were misleading. Cuba had a modest industrial class. Non-sugar industry was making significant strides. Industrial workers were relatively numerous. Cuba was, moreover, urbanized. Since the 1930s, a majority of the population has lived in urban areas. During the 14 years between World War II and the revolution, the middle class expanded notably.

Nonetheless, sugar lay at the core of the problems that Cuba faced during the 1950s. The industry was still the most important depository of domestic and foreign capital investments. Sugar encumbered the creation of employment, and the harvest cycle accounted for high levels of unemployment and underemployment. Without diversification, jobs would not be created and living standards raised. During the 1950s, a consensus was emerging on the need to overcome the centrality of sugar. Moving Cuba from classic dependence on sugar to a new form of dependent capitalism, however, required a realignment of domestic actors, a new role for the state, and a restructuring of Cuba-U.S. relations.

Classic Dependence in Crisis

By the end of the nineteenth century, the United States was the principal market for Cuban sugar exports. While still a colony of Spain, Cuba became commercially dependent on the United States. During the early 1890s, Cuba-U.S. trade took place under terms of reciprocity. In 1894, U.S. preferential tariffs propelled the production of a 1-million-ton harvest. The independence war of 1895–1898, however, devastated the sugar industry: harvests averaged about 215,000 tons a year.[1] In 1898, when U.S. intervention brought the war to an end, the countryside lay in ruins. Most sugar mills were destroyed or inoperative, land lay fallow, and a large segment of the rural population was displaced or decimated. The planter class was bankrupt and lacked the capital to rebuild the mills and replant the cane fields.

The Reciprocity Treaty of 1903 revived the ravaged sugar industry and enabled a seventeenfold expansion between 1900 and 1925.[2] Cuban sugar received a 20 percent tariff reduction in the United States, and U.S. goods 20 to 40 percent in Cuba. The sugar industry attracted substantial foreign investment. By 1925, U.S. capital totaled $750 million, owned 41 percent of all mills, and controlled 60 percent of the harvest. Reciprocity consolidated a mode of development based on monoculture and large landholdings. Prospective sugar profits deflected investment from other sectors and promoted cane cultivation. Nonetheless, under the terms of the 1903 treaty, the sugar industry recovered and, with it, the Cuban economy. By the mid-1920s, Cuban capital was the junior partner, owning less than one-third of all mills and producing less than one-fifth of the harvest.[3] Thus, sugar expansion did not primarily result in national capital accumulation.

After 1925, when world sugar production exceeded demand and prices fell, crisis overcame the sugar industry. The Great Depression further deepened the sugar downturn. Between 1926 and 1940, Cuban sugar output declined more than 50 percent. During the 1940s, higher prices and growing demand resulted in production increases to the levels of 1925 (about 5 million tons a year). During the 1950s, sugar harvests increased slightly.[4] By then, however, population growth had undercut their value. During the 1920s, Cuba had produced about one ton of sugar per person; by the 1950s, the proportion was .86 ton.[5] Not surprisingly, annual per capita income reflected the vagaries of sugar: declining 2.5 percent during the Great Depression, increasing approximately 2.6 percent during the 1940s, and barely moving during the 1950s.[6] Moreover, market conditions for sugar exports were increasingly unfavorable: Cuba was losing ground in the U.S. market and becoming more dependent on the stagnant international market.[7] Nonetheless, throughout the post-1925 period, sugar accounted for about 80 percent of all exports.[8]

The openness of the Cuban economy further aggravated the consequences of sugar monoculture. Between 1945 and 1958, foreign trade averaged 54.8 percent of Gross National Product (GNP).[9] Total trade per capita was among the highest in the world.[10] Relatively declining prices and stagnant markets weakened Cuban terms of trade. Between 1916 and 1925, per capita exports and imports averaged 121 and 90 pesos a year, respectively; during the 1950s, 113 and 103 current pesos. Between 1954 and 1958, the volume of exports and imports grew only 38 and 18 percent in comparison with their 1921–1925 levels.[11] From the late 1920s to the late 1940s, Cuban export prices rose 66 percent; import prices, 85 percent. After World War II, Cuba imported nearly 5 percent less in value and 15 percent less in volume than it had before the depression.[12] The continued primacy of sugar augured further deterioration in the terms of trade.

By the 1950s, Cuba evidently had to turn elsewhere—not necessarily abandoning sugar—in order to resume economic growth. In 1956, the National Bank rendered an ominous report on the consequences for living standards if dependence on sugar were to continue. In 1955, Cuba would have needed a sugar harvest of more than 7 million tons to have maintained 1947 standards of living. A 2 percent yearly increase would have required nearly 9 million tons. The 1955 harvest was less than 5 million tons. By 1965, a harvest of more than 8 million tons would be needed for standards of living to maintain their 1947 levels. To improve them at the 2 percent rate, production would have to be almost 13 million tons.[13] Even if attainable, such harvests were unrealistic without market outlets. The sugar sector had clearly ceased to be the motor for growth.

The centrality of sugar underscored a structural crisis of economic stagnation. Cuba depended on sugar for its export earnings. Sugar cane was planted on well over 50 percent of the land under cultivation.[14] The sugar sector produced about half of all agricultural output and one-third that of industry.[15] It employed 23 percent of the labor force and generated 28 percent of GNP.[16] Since the late 1920s, however, an important change had occurred. U.S. capital had partially withdrawn from the slackening sugar sector and Cubans had gained majority interests. By the early 1950s, Cuban capital controlled 71 percent of all mills and 56 percent of total production.[17] National ownership did not make the movement toward diversification easier, however. Especially during wartime, sugar still offered substantial—albeit relatively falling—profits.

Breaking the sugar conundrum required social initiative, political action, and national vision. The reciprocity mentality was all-pervasive among *hacendados* (mill owners) and most *colonos* (cane growers). Safeguarding the Cuban quota in the U.S. market was their priority. Sugar interests resisted reforms aimed at protecting the Cuban market for national industry for fear that the United States would lessen the preferential treatment of sugar. Moreover, wars tempered the crisis that befell the sugar industry after the 1920s. World War II, the Korean War, and the Suez Canal crisis helped prolong the status quo. The evidence against sugar was unassailable, but making an argument for diversification was difficult.

Since the 1920s, the National Association of Cuban Industrialists had, nonetheless, engaged in that task. ANIC had a vision for a Cuba beyond sugar and pursued it relentlessly. The industrialists took the initiative to address the mounting dilemma besieging the Cuban economy. ANIC, however, never quite mustered full state support nor forged lasting alliances with other social forces to advance a program of diversification. Political action was rarely concerted or sufficiently determined.

Reformism in the Making, 1927–1958

During the early twentieth century, the reconstruction of the sugar industry came to pass at the expense of Cuban ownership, the diversification of agriculture, and the protection of domestic industry. There was, nonetheless, no viable alternative to sugar. Domestic and international circumstances underscored its importance. Cuban planters lacked the capital to rebuild the sugar industry, let alone diversify the economy. Other sectors paled in comparison to sugar and hence could not have generated the resources to restore the economy. Moreover, the turn of the century was a period of extraordinary export expansion in the world economy. What else but sugar could Cuba produce at comparative advantage? Foreign investors—the only source of available capital—naturally gravitated to the sugar industry, where the prospects of profits were most promising. National capitalist development—an elusive passage in Latin America—did not find a track in early-twentieth-century Cuba.

The consequences of restoring the sugar industry were not easily minimized. U.S. capital promoted economic reconstruction but undermined national control of the economy. Trade reciprocity favored U.S. imports, weakened existing industries, and discouraged new ones. More than 350 Cuban-owned establishments closed their doors early in the twentieth century. Expanded trade, moreover, benefited the Spanish-controlled commercial sector. Unlike the rest of Latin America after independence, Cuba did not expel the Spanish. In 1909, Manuel Rionda, a member of the planter class, wrote: "So the Cubans, the real Cubans, do not own much."[18]

Nonetheless, the early republic witnessed the formation of an industrial class. Administration of the state allowed Cubans their only avenue for social mobility. Political elites enriched themselves through corruption and graft. The expansion of public works and services, the promotion of state-development projects, and the establishment of new government agencies also resulted in national capital accumulation. Remnants of the planter class and the Cuban-born sons of Spanish immigrants likewise contributed to the formation of an industrial class. World War I fostered an unprecedented sugar boom and some import-substitution industrialization. Between 1914 and 1920, the value of exports increased four and a half times while that of imports nearly quadrupled.[19] The sugar windfall generated some opportunities for local capital to invest, borrow, and expand the nonsugar economy.[20]

National industry modestly increased its share of the Cuban market. In 1912, consumer goods represented 70 percent of total imports; by 1927, 65 percent.[21] Between 1925 and 1929, local industry produced around 40 to 45 percent of all consumer goods in the domestic market.[22] Without doubt, imports had constituted a much larger proportion of domestic consumption earlier in the century.[23] By the mid-1920s, more than one thousand enterprises were under Cuban ownership.[24] The growth of local industry was not wholly Cuban, however; nonexistent in 1911, U.S. investments in manufacturing amounted to $40 million in 1924–1925.[25]

In 1923, the National Association of Cuban Industrialists was founded. Under the slogan "For the Regeneration of Cuba," the industrialists joined other groups in calling for protection of national industry, repeal of the Platt Amendment, honesty in government, limited presidential terms, and curtailment of foreign land ownership. The reform agenda clearly countered the tenor of Cuba-U.S. relations and the free-trade interests of the sugar industry. In 1925, the election of Gerardo Machado gave reformers cause to rejoice. A general in the Liberation Army and a member of the political elite, the new president was seen to represent the interests of the industrial class. His wealth, considerable and diverse, was a product of the structure of Cuban politics. Partially and briefly, Machado addressed the reformist program. He raised the possibility of revoking the Platt Amendment with the U.S. government and suggested the desirability of revising the Reciprocity Treaty of 1903. The United States responded firmly on the mutual benefits of the Platt Amendment and evasively, if on balance negatively, on treaty review.

Machado obliquely circumvented the issue of sovereignty that had prompted him to question the Platt Amendment. He acted, however, on the matter of protectionism. In 1927, the Customs-Tariff Law established unprecedented protection for national industry. Duties on capital and raw materials imports were lowered and those on many consumer durables and nondurables raised. Existing Cuban industries such as shoe manufacturing, toiletries, furniture, breweries, distilleries, tanneries, dairy, and food processing were protected and new ones were encouraged.[26] For the first time, the state supported economic interests outside the sugar sector. The industrialists wholeheartedly applauded the efforts of the Machado administration.

The reformist movement viewed the sugar industry with profound ambivalence. That the mills were largely foreign-owned, had extensive landholdings, paid *colonos* low cane prices and charged them high land rents spurred the reformist call to regulate the sugar industry and enact land reform.[27] Although some reformers called for the abandonment of sugar, most recognized that only sugar could generate the capital needed for economic diversification.[28] They were, nonetheless, unwilling to accept its immutability and emphasized market diversification for traditional exports as well as outreach to new markets for new products. The industrialists identified Colombia, Panama, Honduras, and Guatemala as prospective markets for Cuban exports such as paints, cordage, ready-made clothing, shoes, and rum.[29] In 1929, Gustavo Gutiérrez, a key figure in the reform movement, succinctly expressed its leitmotif: "Cuba's economic interests should be the basis of our foreign policy."[30] In 1936, Chamber of Commerce leader Luis Machado invoked its quintessence:

> Half a century ago, Cuba's principal challenge was political. Three generations of Cubans struggled and died for the freedom, sovereignty, and independence of our people. Our generation's challenge is economic and social. If our parents forged an independent Cuba, we have to make our country, not only wealthy, but Cuban.[31]

The Customs-Tariff Law of 1927 achieved measured success. Between 1925 and 1933, food products declined from 35 to 29 percent of total imports. Coffee and corn imports were virtually eliminated; meat and lard fell 84 percent; dairy products, 91 percent; potatoes, 86 percent; and rice, 39 percent.[32] Overall trade, however, decreased sharply: exports, 76 percent, imports, 86 percent.[33] The Great Depression further aggravated the deterioration of export earnings and import levels after the mid-1920s sugar crisis. The 90 peso loss (45 percent) in national income per capita, though, was not as steep as that in total trade.[34] Undoubtedly, lower consumption accounted for an important share in import reduction, but had the Customs-Tariff Law not wrought modest achievements, consumption levels would have been even more impoverished.

The Great Depression, nonetheless, undermined the efforts to diversify. Stagnant markets and falling prices reinforced the sugar conundrum. The first step toward economic recovery was improving the situation of sugar. In 1934, Cuba and the United States signed a new reciprocity treaty more favorable to Cuban sugar. The U.S. Sugar Act of 1934, moreover, established a quota system for domestic and foreign producers that increased the share of Cuban sugar. Benefits, however, were relative only to the early 1930s, when Cuba was losing an average of 5 percent a year in the U.S. market and paying the steepest tariffs since the 1890s.[35]

Between 1933 and 1940, Cuban exports to the United States rose 84 percent. The new treaty and a 1939 amendment, however, undermined the Customs-Tariff Law. Broader tariff reductions on U.S. products buoyed the already privileged U.S. position in the Cuban market. U.S. exports to Cuba more than doubled, and the United States increased its share of Cuban imports from 54 to 77 percent.[36] Moreover, U.S. manufacturing investment continued to grow. Between 1929 and 1943, U.S. industrial capital rose from $45 million to $65 million. During the same period, total U.S. investment declined from $859 million to $567 million.[37] Contracting sugar markets and the Great Depression had considerably weakened Cuban export capabilities. The new treaty and the U.S. quota system enabled Cuba to recover. Reciprocity and quotas, however, favored sugar against the reformist program of protectionism and diversification.

The outbreak of World War II further enhanced the prospect of sugar and encouraged a short-term vision. In 1950, the World Bank noted that notwithstanding the shortages that should have encouraged it, diversification "appeared to regress" during the war.[38] Overall domestic consumption declined. Between the 1920s and the 1940s, the value of consumer goods imports fell from 147 to 125 million pesos a year. Domestic production—only 40 to 45 percent of total production—would have had to grow at the unlikely pace of 108 to 123 percent to cover the import deficit in relation to the 1920s.[39] By the late 1940s, per capita consumption of a variety of products declined in comparison with the mid-1920s. While population grew 41 percent, consumption of rice expanded 22 percent, wheat flour 38 percent, potatoes 5 percent, coffee 29 percent, legumes 40 percent, cotton textiles 8 percent.[40] Moreover, the average value of new capital goods was actually higher in the 1920s at 37 million pesos than in the 1940s at 23 million pesos.[41] The sugar industry had ceased to expand and other sectors did not take up the slack in capital investments. World War II did not advance economic diversification.

The industrialists were well aware of missed opportunities. In 1944, ANIC invited all major business associations and trade unions to an economic conference. The meeting outlined a program of protection of national industry, creation of a national bank and merchant marine, revision of extant commercial agreements, and other measures to promote "a broader development of all our productive sectors in view of Cuba's foremost interests."[42] The conference focused on the role of state policy in promoting domestic industry and advocated duty, tax, and other concessions for existing and new industries. The industrialists emphasized how illy prepared Cuba was to face the lower sugar prices of the postwar period. Two years later, they continued to sound the conference's knell:

> In order to obtain our economic independence, it is imperative that we carry out an integral reform of our economic system, which is a colonial one, based on producing raw materials which we sell in only one market at a price and conditions imposed by the buyer.[43]

In 1948, ANIC and the Chamber of Commerce organized another conference, which constituted an exhaustive expression of reformism and covered a broad range of topics: labor-management relations, social security, international trade, fiscal and monetary policies, credit institutions, civil service, and merchant marine. Participants unanimously agreed on most matters, especially on the imperative of greater productivity. Not all, however, concurred on the focal issue of trade. The Sugar Mill Owners' Association and Sugar Cane Growers' Association did not sign the final document. Sugar producers adduced they could not endorse it because their visit to Washington, D.C., for the Department of Agriculture audiences on the 1949 sugar quotas had prevented them from fully participating in the conference. Their procedural rationale notwithstanding, *hacendados* and *colonos* failed to gloss over their disagreement with nonsugar interests over trade policy.

The final document of the conference challenged the bilateral premise underpinning the island's trade relations. Maintaining the Cuban share of the U.S. market did not preempt seeking new markets for sugar. Trade policy was a national endeavor that required private and public coordination. The *clases económicas* as a whole had high stakes in the sugar industry. All sectors, not just the sugar interests, were entitled to a voice in foreign trade agreements, and the conference asserted:

> Bilateralism in trade policy . . . is generally contrary to the national interest and should be substituted by multilateralism. . . . The determination of international commercial policy is a matter of national interest. As such, it corresponds, in the first place, to the public sector, but its elaboration likewise requires input from all economic quarters. . . . Cuba's international trade policy has to be conceived and formulated as an organic whole which harmoniously connects external and internal factors to promote the national interest.[44]

The 1948 conference recognized the primacy of sugar in the Cuban economy. Unlike some reformers during the 1920s, none of the participants suggested phasing out sugar production. Two years before the World Bank commission would note that the problem in Cuba was not too much cane but too little of everything else; the reformers arrived at the same conclusion.[45] Furthermore, they advocated the modernization of the sugar industry. National capital was needed to diversify the economy, and only sugar could generate it. Reformers in the 1940s also aimed to expand markets for Cuban consumer goods in Central America and the Caribbean. The final document noted:

> Export industries are the basic element in the Cuban economy. Nevertheless, their present structure does not meet our needs. The development of a strong economy to satisfy fully domestic consumption and diversify exports is a national imperative if we are to provide full and stable employment . . . We have so many underutilized resources that high export production output is perfectly compatible with industrial and agricultural diversification.[46]

By the 1950s, reformism had attained some successes. A consensus on the interdependence of the sugar industry, agricultural diversification, and import-substitution industrialization appeared to be emerging. Inauspicious market conditions for Cuban sugar seemed to nudge *hacendados* and *colonos* toward considering some degree of change. The realization that Cuba would never again fully regain the lost ground in the U.S. market was particularly persuasive.[47] In 1955, the government unveiled the National Program for Economic Action, which incorporated three decades of reformist prescriptions. With a sense of urgency, the program emphasized that the status quo had ended:

> Cuba cannot continue to depend almost exclusively on sugar to sustain its population, nor wait for solutions through preferential treatment from the United States. . . . The equilibrium between the levels of sugar production and population has been broken. . . . [I]f we do not structure and orient our economy to secure a just and adequate standard of living for our people, unfortunate days await us.[48]

Nonsugar industry was growing at an annual rate of nearly 7 percent.[49] There were more than 2,300 nonsugar industrial enterprises, and Cubans owned the majority. Since the 1930s, imports had undergone a substantial transformation. By 1956, consumer goods had declined to 36 percent of total imports. Capital and intermediate imports had increased to 64 percent.[50] Capital investments in the nonsugar sector were mounting. After having risen during the 1940s, the proportion of food imports to total imports fell in the 1950s.[51] In 1950, the National Bank started operations and other credit institutions were also established. Credit to the nonsugar sector was increasing, sugar credit declining.[52] While still meager, public and private support for research and development was modestly improving.

Progress was slow and uneven, however, Although rates rose through the 1950s, the structure of investment had yet to change significantly enough to support economic transformation. The *clases económicas* enjoyed a favorable climate as wages fell in proportion to national income.[53] Nonetheless, Cuban capital continued to prefer real estate, U.S. bank deposits, U.S. stock and securities purchases, and idle bank balances in Cuba over national industry and agriculture. Cuban nationals had over $300 million in short-term assets and long-term investments in the United States.[54] Agriculture was especially neglected. Between 1951 and 1958, capital goods imports for the sugar industry, nonsugar industry, and agriculture went up 73 percent, 90 percent, and 15 percent, respectively. When the 1953 recession forced a reduction in total imports, agriculture suffered a steep 44 percent loss. In contrast, capital goods imports for sugar and nonsugar industries declined about 20 percent.[55]

Moreover, deficit spending depleted the reserves accumulated during the 1940s. By 1958, monetary reserves had dwindled to 100 million pesos from 571 million in 1952.[56] An increasingly unfavorable balance of payments and international terms of exchange heightened the crisis. During the 1950s, Cuba amassed a 400-million-peso deficit with the United States.[57] Terms of trade were, moreover, rapidly declining relative to the late 1940s, when Cuba had accumulated nearly 1.4 billion pesos in trade surpluses. Between 1950 and 1958, surpluses plummeted to about 367 million pesos.[58] Indeed, the sugar equilibrium was broken.

Nonetheless, the force of sugar still prevailed. The sugar sector was extraordinarily reluctant to break the syndrome of reciprocity. Breaking from the trodden paths—even if increasingly dead-ended—would not come naturally to the sugar industry. In 1955, *hacendados* and industrialists engaged in a fierce public debate about trade with the United States and protection of national industry.[59] Mill owners remained adamant on the safeguards required to secure the U.S. market for Cuban sugar and balked at most efforts to protect the national market. Without reciprocity, they contended, the United States would increase tariffs on Cuban sugar. And where would Cuba be without sugar?

The industrialists argued vehemently that diversification would change only the composition of Cuba-U.S. trade, not its overall amount. The sugar sector continued to insist on the identity of its interests with those of the nation. The Suez Canal crisis accentuated the flawed syndrome of sugar. Higher prices for raw sugar motivated mill owners to reduce the production of molasses, alcohol, and other by-products that were pivotal for the diversification drive. Raw sugar was more profitable, but its by-products more ably promoted the long-term national interest. The industrialists reiterated their long-standing position:

> If the Cuban economy functioned without the limitations imposed on it by foreign interests, it would be a clear sign that we have achieved our economic sovereignty. . . . Everything that is contrary to our economic progress, to the industrialization and diversification of our economy, to the right of all the citizens of this land to a job and sufficient income, that is the enemy against which the industrialists struggle.[60]

At the end of the 1950s, the sugar and nonsugar sectors of the *clases económicas* were evidently at an impasse. What role had the state played in relation to the central issues and actors in the Cuban economy?

State and Society

Before the 1920s, the Cuban state had largely refrained from intervention in the economy. Only during the Liberal administration of José Miguel Gómez (1908–1912) did the state initiate development projects. The collapse of international markets in the mid-1920s, however, impelled the Machado administration to regulate the sugar industry. With the Verdeja Act of 1926, the government restricted Cuban harvests in the hope of raising sugar prices. Quotas were assigned to each mill based on past production, acreage under cultivation, and the number of *colonos* providing it with cane. The length of the harvest was shortened and new plantings suspended.

Nonetheless, regulation did not arrest the drop of sugar prices. After two restricted harvests in 1926 and 1927, the 1928 price of sugar fell more than 40 percent relative to that of 1924. Moreover, while Cuba reduced its harvest by more than 1 million tons, world production expanded by more than 2 million.[61] Even so, the policy of harvest restriction signaled the modus operandi of state intervention in the face of stagnant world demand and growing market competition. For more than two decades, the state intermittently pursued a policy of restriction even though the benefits to Cuba were dubious.

The Sugar Stabilization Institute (1931) and the Sugar Coordination Act (1937) institutionalized state regulation of the sugar sector. The institute incorporated representatives of mill owners, cane growers, the government, and the unions, and was responsible for enforcing sugar regulations and conducting international negotiations. The 1937 act established a grinding quota so that crop restrictions would not unduly burden *colonos* and the smaller mills. The act also secured the right of *colonos* to permanent land tenure: as long as quotas were met and rental payments made, cane growers could not be evicted. However, land tenure was made contingent upon the fulfillment of cane quotas. Because *colonos* had no incentive to cultivate other crops, the act actually deterred agricultural diversification.[62] Sugar was too central to the national well-being and its domestic constituencies too powerful for the state to abstain from intervention in a crisis. Regulation, however, entrenched the status quo.

Only World War II helped Cuba to rebound from the throes of the Great Depression. War prosperity poignantly underscored the vulnerability of the Cuban economy to a volatile market. War allowed Cuba to prosper; peace augured shrinking international markets and reduced quotas in the U.S. market. The industrialists notwithstanding, the Cuban government continued to insist on the logic of reciprocity. During the war, Cuba sold sugar to the United States below market prices. Subsequently, the Cuban government and the sugar sector expected the United States to remember their wartime cooperation in curbing sugar prices and expressed "hopes for equity and reciprocity" in the postwar order.[63] For a brief moment, it seemed such hopes might indeed be realized.

In 1946, the United States proposed that Cuba sell the 1946 and 1947 harvests at three to four cents below the world market price. The Sugar Mill Owners Association supported the U.S. proposal; the Cuban government, the *colonos*, the unions, nonsugar interests, and many *hacendados* rejected it. Their demands included the sale of one harvest at a time; a larger share for Cuban sugar in the U.S. market; larger quotas for molasses, alcohol, and refined sugar; and, most important, a "guarantee clause" that sugar prices would rise in tandem with those of U.S. exports to Cuba. Some *hacendados* and *colonos* even hinted sugar shipments to the United States might be suspended if these demands were not addressed. The immediate outcome of this uncharacteristic confrontation was the separate and more beneficial negotiation for Cuba of the 1946 and the 1947 harvests and the 1947 inclusion of the "guarantee clause."[64]

Shortly thereafter, however, the Sugar Act of 1948 precipitated a debate in the U.S. Congress that underscored the uniqueness of the 1946 encounter. One of its appended clauses stipulated the reduction of the sugar quota of any country that did not treat U.S. citizens and their interests "equitably," a thinly veiled exigency against Cuba because Cuban citizens owed U.S. concerns $8 million. Not coincidentally, the clause followed upon the more assertive Cuban negotiations for the 1946 and 1947 harvests. In response, Cuba pursued the inclusion of a provision in the Río Treaty of 1947 against economic coercion. The Río conference failed to pass the Cuban motion, the U.S. Congress eventually repealed the disputed clause to the Sugar Act of 1948, and the Cuban government subsequently refrained from such an assertive negotiating stance with the United States.[65] Cuba had, once more, learned the limits of its sovereignty. The 1946 negotiations and the 1947 controversy over the U.S. Sugar Act highlighted the magnitude of the obstacles that a transition to a new form of dependent development faced in Cuba. It was impossible without Cuba's redefining its relationship to the United States, yet, the United States insisted on a singular and intimate relationship.

Defending the preferential treatment of sugar often undercut domestic efforts to diversify the economy. Rice agriculture exemplified the obstacles confronting the nonsugar sector. The Customs-Tariff Law of 1927 had established protective tariffs to stimulate rice cultivation. The government had distributed seeds and disseminated technical information in rural areas to encourage domestic production. At the time, Cuba purchased almost all rice imports from the Far East. During the 1930s, the revised reciprocity treaty accorded U.S. rice exports to Cuba a 50 percent tariff reduction over those from Asia. Subsequently, Far East imports all but disappeared from the Cuban market. By the end of the decade, the United States was supplying most Cuban rice imports.[66] Nonetheless, between 1936 and 1941, total rice imports declined by 10 percent while domestic production nearly doubled. U.S. and Cuban rice was now satisfying domestic consumption[67] of the single most important item in the Cuban diet.[68]

Cuban producers fought an uphill and ultimately losing battle to dominate their own market. Between 1941 and 1958, rice agriculture expanded almost twentyfold and rice imports increased less than 2 percent. A separate look at the 1940s and the 1950s reveals another story. During the 1940s, rice imports actually grew about 50 percent. By 1955–1956, domestic production had risen to satisfy 52 percent of consumption, and imports fell below their 1941 level. U.S. rice growers systematically protested the decline in Cuban imports. The U.S. Department of Agriculture conveyed their concern to the Cuban government and implied that the sugar quota might be reduced.

Cuban sugar and commercial interests lobbied to defend the bastion of sugar. The newly constituted state banks failed to extend credits to support the extension of rice cultivation. Unnecessary rice imports were authorized to assuage U.S. rice exporters. Cuban rice growers protested to no avail. Then, under the Sugar Act of 1956, the United States formally secured a Cuban commitment to purchase rice in exchange for continued preferential treatment of sugar. Cuba purchased about 75 percent of all U.S. rice exports. Between 1955 and 1959, domestic production grew about 10 percent and rice imports more than 40 percent. The proportion of national consumption satisfied by Cuban producers receded to about 45 to 47 percent. After a substantial loss, U.S. exports again surpassed their 1941 levels.[69] The Cuban rice industry disclosed the entrapment of the state in the imperatives of sugar production. Because of U.S. and domestic opposition, state banks did not support rice production and, consequently, the state failed to promote national interests. Significantly, the controversy over sugar and rice happened after the government had announced the reformist-oriented National Program for Economic Action.

In many ways, the ambience of the 1950s was propitious to implementation of the reform program. The Batista administration enacted a protective tariff that favored the purchase of raw materials and capital goods and curtailed consumer goods imports more strictly than had the Customs-Tariff Law of 1927.[70] New measures to guarantee foreign investments such as tax exemptions and more liberal terms for capital remittances were also passed.[71] The rates of domestic and foreign investments were increasing. In 1957–1958, domestic investments had grown nearly 50 percent relative to 1950–1951.[72] Between 1956 and 1960, U.S. capital had projected a $205 million influx of new, nonsugar investments, a 20 percent increase over its total in Cuba.[73] Three of the constituent factors of dependent development—a more activist state, national capital, and foreign investments—were potentially available, but the confrontation over the U.S. Sugar Act in 1956 underscored the chasm between possibilities and reality.

Cuban society, nonetheless, harbored the social forces that might have served the long-delayed program of reform. Among the *clases económicas,* the industrialists were a growing voice for economic transformation. The relatively large middle class—perhaps about one-third of the population—also represented a constituency against the status quo.[74] The middle sectors included 179,571 professionals, managers, and executives, nearly 10 percent of the economically active population.[75] The 203 professional associations were middle-class organizations.[76] The middle class, moreover, was expanding.[77] Crucial sectors of the unionized working class, especially in Havana and in the more modern industrial enterprises, were also potential supporters of renewing dependent capitalism. During the 1950s, the contours for the transformation of classic dependence were present. There was a sense of urgency about that transformation: the primacy of sugar preempted improvements in standards of living. A growing sense of insufficiency permeated Cuban society.

Standards of Living

The problem of employment lay at the core of the old Cuba (see Table 1.1). In the mid-1950s, one-third of the labor force did not hold full-time jobs. During the dead season, overt unemployment rose to 20.7 percent; underemployment averaged 13.8 percent throughout the year. About 20 percent of the economically active population (EAP) worked in industry, 40 percent in agriculture, and 30 percent in commerce and services. Unemployment levels had remained unchanged since the early 1940s.[78] Urban-rural disparities highlighted the magnitude of the problem: 71 percent of the urban labor force and 64.3 of the rural had full-time jobs year-round.[79] Interestingly, rural Cubans identified jobs—not access to land—as their foremost need: in a 1956–1957 survey 75 percent responded employment opportunities were most important in improving their living conditions. Nearly 69 percent looked to the state for solutions.[80]

Educational levels underscored the uneven modernization of Cuban society (see Table 1.2). School enrollment among 5- to 14-year-olds expanded rapidly until the mid-1920s, when it began to decline. In 1953, the census registered significant improvement without yet matching the

Table 1.1. Employment, Unemployment, and Underemployment in Cuba, 1950s (in percentages)

	Urban	Rural	National	
	1953	1953	1953	1956–1957
Population	57.0	43.0	100.0	100.0
Unemployment	9.7	6.6	8.4	16.4
Underemployment[a]	17.1	16.5	16.9	13.8
Full-time employment[b]	71.0	64.3	68.4	65.3

Sources: República de Cuba, *Censos de población, viviendas y electoral: informe general* (*enero 28 de 1953*) (Havana: P. Fernández, 1955), pp. 153, 176; Informe de la Comisión Coordinadora de la Investigación del Empleo, Sub-Empleo y Desempleo, *Resultados de la Encuesta sobre Empleo, Sub-Empleo y Desempleo en Cuba* (*mayo 1956–abril 1957*) (Havana, January 1958), pp. 41, 50.

[a] 1953 underemployment = 29 weeks or less a year; 1956–1957 underemployment = less than 29 hours a week + without pay for a relative.

[b] 1953 full-time employment = more than 50 weeks a year; 1956–1957 full-time employment = 40 or more hours a week.

levels of the 1920s.[81] Literacy rates revealed a similar pattern: improvements through the 1920s, decline during the 1930s, and subsequent recovery.[82] During the 1950s, Cuban literacy rates were the fourth highest in Latin America after Argentina, Chile, and Costa Rica. Cuba, however, placed twelfth among Latin American countries in school enrollment among 5- to 24-year-olds, even though it had the highest educational expenditures relative to national income.[83] Urban-rural differences highlighted the unevenness of these levels. Among rural Cubans, illiteracy was nearly four times higher and school enrollment of 5- to 14-year-olds less than half. Overall, urban Cubans reached signifi-

Table 1.2. Educational Levels in Cuba, 1953 (in percentages)

	Urban	Rural	National
Illiteracy[a]	11.6	41.7	23.6
School enrollment[b]	69.0	34.9	51.6
Third grade or less[c]	44.7	83.3	60.4
High school/vocational graduates[c]	5.8	0.4	3.5
University graduates[c]	1.8	0.06	1.1

Source: República de Cuba, *Censo de población, viviendas y electoral: informe general* (*enero 28 de 1953*) (Havana: P. Fernández, 1955), pp. 99, 131, 143.

[a] Population 10 years and older.

[b] Population 5 to 14 years old.

[c] Population 6 years and older.

cantly higher educational attainments.[84] After jobs, education was the most common demand among rural workers.[85]

The health profile of Cuba manifested similar inequalities. Life expectancy at 58.8 years, crude death rates at 6.4 per 1,000 persons, and infant mortality at 37.6 per 1,000 were among the best in Latin America. As in Argentina and Chile, two of the top three causes of death were decidedly modern: cardiovascular diseases and malignant tumors. During the 1950s, most other Latin Americans succumbed to diseases of poverty such as digestive-system complications, infancy-related illnesses, and respiratory disorders. The doctor-to-population ratio was second highest and the hospital beds-to-population ratio ranked among the top ten in Latin America.[86] Yet in 1950, the World Bank observed: "Disease is not a serious problem in Cuba, *but health is.*" An overwhelming majority of rural children suffered from intestinal parasites. About half of all Cubans registered some degree of undernourishment.[87] Rural workers had a 1,000-calorie daily deficit and were 16 percent under average height and weight.[88] Sixty percent of physicians, 62 percent of dentists, and 80 percent of hospital beds were in Havana. There was only one hospital in rural Cuba.[89] In 1956–1957, four out of five rural workers received medical attention only if they paid for it, and hence most had no access to health care.[90]

Marked urban-rural differences also characterized housing conditions. Nationally, 43 percent of all housing units lacked running water, 23 percent an inside or outside toilet, 56 percent a bath or a shower, 75 percent a refrigerator, while nearly 60 percent had electricity. More than 50 percent of the units were constructed of solid materials; 15 percent were classified in poor condition. Urban Cubans were more likely to live in dwellings with electricity (87 percent), a refrigerator (38 percent), running water (82 percent), an inside or outside toilet (95 percent). Most of their homes were built of solid materials (86 percent) and rated in good or fair condition (91 percent). Most rural Cubans lived in housing without running water (85 percent), an inside or outside toilet (54 percent), electricity (93 percent), a refrigerator (96 percent). Their homes were more frequently in poor conditions (26 percent) and built with inferior materials (91 percent). There were fewer than 150,000 radios and 4,000 television sets in rural areas; urban Cuba had nearly 475,000 and more than 75,000.[91]

The chasm between Havana and the rest of the island was greater than that between urban and rural Cubans. Twenty-six percent of the population lived in Havana province, most in urban areas. Unemployment and underemployment afflicted *habaneros* less than other Cubans. Havana was less dependent on agriculture and had nearly 50 percent of all industries, including eight of the fourteen plants with more than five hundred workers. At 9.2 percent, Havana illiteracy was well below the national and urban averages. School enrollment for the 5- to 14-year-

old population was 74 percent. *Habaneros* were more likely to go beyond the third grade and attain some level of intermediate primary
education (52 percent). More graduates from high school (60 percent),
vocational school (50 percent), and higher education (70 percent) lived
in Havana.[92] (see Table 1.3).

Income distribution trends dramatically underscored the primacy of
Havana province. Between 1952 and 1958, total national wages barely

Table 1.3. Selected Indicators, Province of Havana and the Rest of
Cuba, 1953 (in percentages)

	Havana	Others
Total population	26.3	73.7
Urban	91.0	44.7
Total housing	32.0	68.0
Illiteracy[a]	9.2	33.2
School enrollment[b]	82.4	45.0
Third grade or less[c]	38.0	69.0
High school/vocational graduates[c]	6.9	2.2
University graduates[c]	2.7	0.4
Unemployment		
1953	9.2	8.1
1956–1957	11.8	18.4
EAP[d]		
Agriculture	10.4	55.4
Industry	20.6	14.8
Commerce/services	53.0	22.4
Total wage bill		
1952	53.0	47.0
1958	64.0	36.0
Absolute wages[e]		
1952	379	337
1958	463	260

Sources: República de Cuba, *Censos de población, viviendas y electoral: informe general (enero
28 de 1953)* (Havana: P. Fernández, 1955), pp. 21, 99–100, 131–32, 143–144, 153, 157,
185–186, 208; Informe de la Comisión Coordinadora de la Investigación del Empleo, Sub-
Empleo y Desempleo, *Resultados de la Encuesta sobre Empleo, Sub-Empleo y Desempleo en
Cuba (mayo 1956–abril 1957)* (Havana: January 1958), pp. 41, 68; Banco Nacional de
Cuba, *Memoria, 1958–1959* (Havana: Editorial Lex, 1960), pp. 151–153; Raúl Cepero
Bonilla, *Escritos económicos* (Havana: Editorial de Ciencias Sociales, 1983), pp. 416–417.
[a] Population 10 years and older.
[b] Population 5 to 14 years old.
[c] Population 6 years and older.
[d] Economically Active Population.
[e] Million pesos. Excludes sugar agricultural workers and only partially includes other
agricultural workers.

grew, yet Havana increased its share from 53 percent to 64 percent; that of each of the other five provinces declined or remained the same. The regression in Las Villas (from 10 percent to 8 percent), Camagüey (from 13 percent to 8 percent), and Oriente (from 14 percent to 12 percent) was especially onerous. Matanzas (5 percent) and Pinar del Río (3 percent) roughly maintained their shares of total wages.[93] Construction and nonsugar industry spurred Havana's gains; the sugar-dominated economies of Las Villas, Camagüey, and Oriente accounted for their regression. The second-largest nonsugar industrial sector in Matanzas and a more diversified agriculture in Pinar del Río largely prevented the erosion of their total wages.[94] Havana's principal advantage lay in its disproportionate share of wage earners whose monthly salaries were 75 pesos or more. About 23 percent of all wage earners worked in occupations in which at least half had wages of 75 pesos or more, and 51 percent of those lived in Havana. In contrast, the province had a commensurate share (26 percent) of persons in occupations in which more than half earned less than 75 pesos a month. Intensifying income inequalities accompanied Havana's increasing proportion of total wages.[95]

In 1961, the writer Lino Novás Calvo noted: "Anyone who had known Havana in 1914 and saw it again in 1958 would have been amazed at its progress. No other Latin American country had advanced so far in so short a time."[96] Havana, however, was not Cuba. The capital was quite modern and *habaneros* enjoyed relatively high standards of living. Most were literate, had achieved higher levels of education, had more access to health care, were more likely to be permanently employed, and earned better wages than Cubans in the provinces. Havana, moreover, was undergoing a consumer boom. During the 1950s, Cuba imported an average of 30 million pesos in cars and 45 million pesos in household durables a year. Between 1956 and 1958, the latter jumped to 63 million pesos.[97] While other urban areas received some of these consumer goods, most made life in Havana a bit more comfortable. *Cómprelo a plazos* (buy it on the installment plan) became commonplace in advertising. Some New York and California department stores ran regular advertisements in Havana newspapers.[98] Like the United States during the 1950s, the Cuban capital was on the brink of mass consumerism.

Most Cubans, however, were not *habaneros*. Their progress, especially in rural areas, had been considerably more erratic and uneven. In 1957, the Catholic University Association noted:

> Havana is living an extraordinary prosperity while rural areas, especially wage workers, are living in unbelievably stagnant, miserable, and desperate conditions. . . . It is time that our country cease being the private fiefdom of a few powerful interests. We firmly hope that, in a few years, Cuba will not be the property of a few, but the true homeland of all Cubans.[99]

Women in Prerevolutionary Cuba

Early-twentieth-century Cuba saw the development of an impressive feminist movement. More akin to the U.S. and British movements, Cuban feminism was another expression of relative modernity before 1959. The mostly privileged women who integrated the movement focused on legal reform: the right to vote; laws on marriage, property, and divorce; protection for out-of-wedlock children; and, secondarily, labor legislation.[100] During the 1940s and 1950s, some women's organizations continued to be active, if not on explicitly feminist issues. In 1954, for example, a congress of women focused on the economic problems besieging Cuban families, the health care system, the educational crisis, public morality, the restoration of democracy, and, less prominently, the concerns of working women.[101] Nonetheless, after women gained the right to vote in 1934 and the Constitution of 1940 incorporated most feminist claims, the feminist movement dissipated.

The socioeconomic profile of women underscored the uneven modernization of Cuban society (see Table 1.4). Women constituted a lower proportion (13 percent) of the economically active population in Cuba than elsewhere in Latin America. The economy was not generating sufficient employment for men, let alone women. Sugar cane, moreover, was not based on the traditional forms of agriculture that in many other Latin American countries engaged the labor of women. Still, working women were more likely to have a full-time job than men: 75.9 percent as opposed to 66.8 percent. Commerce and services employed most working women (72.5 percent). Household and personal services accounted for more than one-third of female employment; industry for less than a fifth.[102] Nearly half of the male labor force worked in agriculture. Industry—a fast-growing source of male employment—employed 21 percent of working men.[103]

The overall profile of men and women differed in other ways. Working women were more likely to live in urban areas (78 percent), especially Havana (40 percent). Only about 55 percent of the male labor force worked in urban areas, 28 percent in Havana. Government represented a larger share of the EAP for women (25 percent) than for men (6 percent). Like women in general, working women had higher levels of education than men: 20 percent of women were skilled workers and 16 percent professionals; 18 percent and 3 percent of working men had respectively similar levels. Nonetheless, women earned less than men: 29 percent had monthly salaries of 75 pesos or more; 40 percent of the men received such sums. More than 75 percent of all skilled women earned less than 75 pesos a month, in contrast to less than half the men in the same category. More than 20 percent of professional women had monthly salaries below 75 pesos; about 15 percent of the men did.

Table 1.4. Selected Indicators: Men and Women in Cuba, 1950s
(in percentages)

	Men	*Women*
Unemployment		
1953	9.0	5.8
1956–1957	17.1	11.8
Underemployment[a]		
1953	7.8	14.1
1956–1957	12.8	8.0
EAP[b]		
Agriculture	46.9	5.7
Industry	21.0	19.7
Commerce/services	25.8	72.5
Earnings above 75 pesos a month	39.5	28.9
Illiteracy[c]	25.9	21.2
School enrollment[d]	51.5	51.6
Third grade or less[e]	61.7	59.1
High school/vocational graduates[e]	3.4	3.6
University graduates[e]	1.4	0.7

Sources: República de Cuba, *Censos de población, viviendas y electoral: informe general (enero 28 de 1953)* (Havana: P. Fernández, 1955), pp. 99, 119, 143, 153–154, 185, 195; Informe de la Comisión Coordinadora de la Investigación del Empleo, Sub-Empleo y Desempleo, *Resultados de la Encuesta sobre Empleo, Sub-Empleo y Desempleo en Cuba (mayo 1956–abril 1957)* (Havana: January 1958), pp. 41, 50.

[a] 1953 underemployment = 29 weeks or less a year; 1956–1957 underemployment = less than 29 hours a week + without pay for a relative.

[b] Economically Active Population; women were 13.0 percent of the EAP.

[c] Population 10 years and older.

[d] Population 5 to 14 years old.

[e] Population 6 years and older.

Although 96 percent of the women in services failed to command wages of 75 pesos and above, only 64 percent of the men did. All women engaged in agriculture made below the 75 pesos level, 84 percent of the men did.[104]

The Cuba That Might Have Been

During the 1950s, classic dependence was coming to an end. Sugar still dominated the economy and deterred significant diversification. Nonetheless, the transformation of monoculture appeared to be a matter of time. U.S. capital was once again beginning to flow into Cuba. The government had issued a comprehensive program for economic development. New avenues other than industrial production for the domestic market were opening up. During the 1950s, tourism, especially to the casinos in Havana and the beaches between the capital and Varadero, Matanzas, considerably expanded. In Camagüey, the King Ranch of Texas was introducing modern cattle-raising methods that Cuban cattlemen were beginning to incorporate in transforming the industry. They would probably have claimed a niche in the soon-to-expand fast-food market in the United States. Growing winter vegetables for the U.S. market was also increasing. Cuban firms already excelled in mass media, advertising, and entertainment, some of which linked the Latin American and the U.S. markets. During the 1960s, these industries would have probably flourished in a Cuba not in revolution. Initial explorations had resulted in optimistic expectations about Cuban petroleum deposits. Broadening employment vistas in other sectors might have more easily allowed the modernization of the sugar industry. Without the revolution, Cuba might have taken the path of tropical dependent development—a missing model of dependent capitalist development in Latin America.

The road not taken would have been unlikely to foster national capitalist development and stable representative democracy. Elsewhere in Latin America, dependent development did not turn out to be particularly nationalist nor especially democratic. Quite the contrary. During the 1950s, the Economic Commission for Latin America inspired development programs that, for a variety of reasons, would fail to attain their objectives. Hence, it is unlikely that transformed dependence would have led to national capitalist development in Cuba, as it did not in the rest of Latin America. Tropical dependent development would not have necessarily sustained democracy either. A capitalist take-off required a more favorable business climate than the militance of the Cuban working class provided. Military governments harsher than that of Batista during the 1950s could conceivably have been in the offing.

The Cuban economy, moreover, already presaged situations that would later characterize much of Latin America. That during the 1950s Cuba suffered from deteriorating terms of trade and an unfavorable balance of payments pointed to mounting foreign debt. Migration from Cuba to the United States increased from about 3,000 a year between 1950 and 1954 to over 12,000 between 1955 and 1958.[105] Development patterns were already forcing increasing numbers of Cubans to seek their fortunes elsewhere, largely in the United States; their intensification would have in all likelihood resulted in even more significant migration. Underworld operations proliferated in the decade before the revolution. A 1958 law facilitated international transactions by Cuban banks. The drug trade and a banking sector, not unlike that which later developed in Panama, would likely have flourished in the Cuba that might have been.

That the transition from classic dependence toward some form of dependent capitalist development never happened in Cuba was due in no small part to political factors. The class and state alliances that might have sustained such a transition never quite consolidated, and those that supported the revolution and its radicalization in 1959 did. Structural conditions underscored the impasse of the sugar status quo and pointed to undercurrents of change. Uneven modernization had also created the social forces to sustain movement toward tropical dependent development. Neither proved sufficient to uphold capitalism in Cuba. Between 1902 and 1958, the functioning of politics—in the state and among opposition movements—eventually provided the catalyst for the revolution and the basis for socialism as an option after 1959.

CHAPTER 2

Politics and Society, 1902–1958

> Without workers, there is no sugar.
>
> *Lázaro Peña*
> *Central Organization of Cuban Trade Unions*
> *1940s*

Political factors were crucial in the coming to power of the Cuban Revolution. During the 1890s and 1930s, the United States helped Cuban elites to defuse popular challenges. Immediate successes, however, were not conducive to long-term political stability. Disbanding the *Ejército Libertador* and inaugurating the Cuban Republic under the Platt Amendment did not promote the order the United States and propertied creoles had hoped. After 1902, there were two U.S. military interventions and countless other civilian intromissions. Among Cuban *políticos,* retaining or gaining access to the public treasury was the primary electoral concern. Once in office, enrichment for themselves and their supporters constituted their first consideration. With the emergence of new social and political forces, the 1930s brought the Plattist republic to an end.

The Constitution of 1940 was the compromise that settled the revolutionary struggles of the 1930s. It included the recognition of many social and economic rights as well as protection of civil liberties and private property. Under its charter, representative democracy was reconstituted and three presidents were elected, but new *políticos* and political parties continued the tradition of corruption. In 1952, a military coup preempted the constitution as Fulgencio Batista restored the army to political preeminence. During the 1950s, an opposition movement mobilized the polity and, after two years of armed struggle, succeeded in toppling the dictatorship. By 1959, accumulated societal crises had considerably weakened the forces that could have moderated the revolution. In addition, the dynamics between Batista and the opposition movement enhanced the weight of Fidel Castro, the Rebel Army, and the July 26th Movement in the victory of January 1. Thus, the long-term crisis of political authority and its more immediate expression in the

36

Batista regime rendered Cuba vulnerable to the possibility of social revolution.

Mediated Sovereignty and Fragile Hegemony

On May 20, 1902, the Cuban flag was raised over Morro Castle at the entrance of Havana harbor. Thirty-four years after the Ten Years' War first called for *Cuba libre*, the Cuban Republic came to pass with much poignancy and no small bitterness. In 1898, U.S. intervention had frustrated the *Ejército Libertador* in the final onslaught against Spanish colonialism. Between 1898 and 1902, the United States occupied the island to safeguard order, property, and privilege. For a while, the inauguration of an independent Cuba was very much in doubt. With the inclusion of the Platt Amendment in the constitution, the United States finally agreed to Cuban independence.

Cuba libre was born under circumstances different from those popular *independentismo* had anticipated. The organizations of the 1895 independence movement had virtually no sway in the birth of the republic: the Cuban Revolutionary Party no longer functioned and the U.S. Army disbanded the *Ejército Libertador* in 1899. Two of the three leaders of independence—José Martí and Antonio Maceo—were dead. The third—Máximo Gómez—was old and alone, and had acquiesced in the dissolution of the Liberation Army. In 1901, the constitutional convention faced a rending dilemma: accept the Platt Amendment that so flagrantly constrained national sovereignty or reject it, knowing that without it there would be no Cuban Republic. By a single vote, the conventionists rejected intransigence and settled for mediated sovereignty.

Social disarticulation marked the early republic. The planter class had little choice but to relinquish economic reconstruction to foreign capital and bind its well-being to U.S. investments. The consolidation and expansion of Spanish interests also limited Cuban opportunities in commerce, industry, and the professions. National and racial differences diffused the *clases populares*. Massive immigration—principally from Spain, Haiti, and Jamaica—swelled the ranks of the working classes to satisfy the labor demands of a rapidly expanding sugar industry.[1] Unemployment, underemployment, and depressed wages accompanied the expansion of foreign capital. Labor unrest threatened order, and maintaining an auspicious ambience for capital was the litmus test early republican governments had to pass in order to avoid U.S. intervention. Containing the *clases populares* was the *sine qua non* of mediated sovereignty.

Stable governments were imperative to safeguard a modicum of independence. Nonetheless, the conditions under which the republic was founded undermined the stability needed to sidestep U.S. intervention. Although economic recovery rested on a favorable climate for foreign investments, economic expansion provoked the mobilization of labor. The state, in turn, could not strike compromises with labor similar to those achieved in other Latin American countries at the time.[2] Foreign capital rejected concessions and demanded order. Because foreigners controlled industry and commerce, public office became the exclusive realm of Cubans. Control of the state bureaucracy provided access to resources inaccessible elsewhere. Thus, presidential reelections became the focus of contention as incumbents were loath to relinquish power.[3]

Politics in Plattist Cuba quickly acquired a pervasive logic. Losers often charged fraud and contested electoral results. In 1906, U.S. intervention led to a three-year occupation and the reorganization of the Rural Guard into a regular army more effectively equipped to safeguard order. The military had high stakes in the orderly conduct of elections and tended to support whoever succeeded in establishing incumbency. Prolonged contestation among opponents brought the threat of U.S. intervention and a situation in which the army, however improbably, might be called upon to defend the nation. The Cuban military clearly lacked the disposition to challenge the United States.[4]

In principle, the political class likewise sought to stave off U.S. intervention. Application of the Platt Amendment was a blatant reminder of the limits of Cuban independence. Nonetheless, appealing to the United States to settle electoral disputes became normal. Elections were not, moreover, the only occasions of U.S. intromission. The United States, for example, insisted there be honesty in public administration, but virtuous management of the public treasury contravened the logic of early republican politics. For the political class, corruption was the unwritten condition of stability. To the United States, malfeasance in office was evidence of the limited ability of Cubans for self-government. In 1921, the United States sent General Enoch Crowder aboard the battleship *Minnesota* on a mission to promote rectitude in the conduct of public affairs. The U.S. delegation departed without much success. The underpinnings of Cuban politics ran counter to the reforms the United States sought to implement under the Platt Amendment.[5]

Movements from various quarters soon challenged the politics of Plattist Cuba. During the 1910s, sugar, tobacco, construction, railroad, and port workers went on strike with relative frequency. In 1914, the Mario García Menocal administration supported the celebration of a labor congress but did not succeed in co-opting the nascent labor movement.[6] The United States opposed two key labor demands: a minimum wage and a Cuban-majority work force in all enterprises. The Cuban legislature never acted on these demands,[7] and repression of strikers and union leaders increased. During World War I, U.S. marines stationed in Oriente were often mobilized to areas of labor unrest. During the 1920s, the Machado administration assassinated militant labor leaders while recognizing the right of maritime workers to unionize in the hope of countering radical labor organizations. Like Menocal, Machado failed to co-opt the union movement.[8] A quiescent working class was increasingly elusive.

During the 1920s, the United States found the terms of its relations with Cuba progressively problematic. Continuous intervention did not beget stable governments capable of maintaining order and defending foreign capital. Presidential successions were almost always moments of turmoil. Actual or threatened military intervention did not bring lasting peace and tranquility. Constant U.S. interference exposed the political class and provoked growing nationalist demands from labor and the reformers. The Platt Amendment subverted the Cuban political elite. A new Cuban policy would, moreover, help to assuage Latin American discontent over U.S. interventions in Mexico and the Caribbean.[9]

The election of Gerardo Machado offered Washington the opportunity to establish a new mode of interaction with Cuban elites. The new president favored Cuban interests without alarming the United States. The Customs-Tariff Law of 1927 incorporated most of the reformist agenda. The construction of the Central Highway impugned the monopoly of foreign-owned and sugar-centered railroads. The advancement of education was significant as schools were built and enrollments increased. At the same time, Machado unrelentingly repressed popular unrest and thereby reassured the U.S. government of his commitment to defend foreign capital. Cuba finally had a government capable of securing social peace without U.S. intervention. Nonetheless, the *machadista* program unraveled.

The sugar crisis after 1925 bode ill for the Machado government. Diversification was stunted, unemployment increased, standards of living fell, and per capita income decreased. With declining revenues, public works were suspended. Many state employees were laid off; others saw their pay reduced 60 percent. Moreover, Machado sought reelection contravening his electoral promise. In 1928, conservatives and liberals— their three-decade rivalry notwithstanding—formed a coalition in support of extending Machado's presidential term. The *clases populares* were challenging the status quo and Machado seemed competent for the task of containing them. Known as *cooperativismo,* the new arrangement signaled a rupture in the pattern of Cuban politics.

Cooperativismo provoked widespread opposition. Demanding university autonomy, students in Havana promoted antigovernment activity. As the economy deteriorated, the working class turned more militant. In March 1930, 200,000 workers went out on strike. As new organizations contested the government, Machado responded with intimidation, harassment, and repression. The ABC Revolutionary Society, the Directorate of University Students (DEU), the Cuban Communist Party (PCC), and the National Confederation of Cuban Workers

(CNOC) countered official repression with violence of their own. The political class itself divided: dissenting members formed the Nationalist Union and led an unsuccessful armed uprising against the *machadato*. With the upsurge of opposition from all quarters, Machado turned more intransigent in the exercise of power.

Until 1933, the United States did not actively intervene in the unfolding crisis. The State Department welcomed *cooperativismo* and the reelection of Machado. Initially, the repression of workers and students did not arouse U.S. concern: Machado was using force to keep the peace, and as long as he was successful, there was no need for U.S. intervention. Nonetheless, the United States was drawn into the maelstrom. Economic depression turned political conflict into social crisis. Early republican politics was running its course, and U.S. intervention seemed unavoidable.

In May 1933, Franklin Delano Roosevelt appointed Assistant Secretary of State Sumner Welles ambassador to Cuba. For nearly seven weeks, Welles mediated between the Machado government and the "responsible" opposition sectors. The more radical anti-Machado organizations like the CNOC, the Communist party, and the Directorate of University Students did not recognize the right of the ambassador to arbitrate; the Nationalist Union and the ABC did. Welles pursued two immediate objectives so as to avoid direct U.S. intervention: removing Machado from office and forging a new consensus among members of the political class, the army, and the "responsible" opposition. In a last attempt to remain in office, Machado resisted the mediation, denounced the United States, and even implied that the Cuban army would fight the marines. The officer corps winced at the prospect and turned against the president. On August 12, 1933, Machado left for the Bahamas.

The mediation was not ultimately responsible for the fall of Machado. A general strike was. The government had met striking bus drivers in Havana with violence. Sympathy strikes followed and over 200,000 workers paralyzed the capital and other cities as labor unrest also spread to the countryside. Allegedly to avoid U.S. intervention, the communists who controlled the CNOC reached a last-minute compromise with Machado and called off the strike but the workers did not respond. After Machado left, Welles ushered Carlos Manuel de Céspedes into the presidency, but his government did not survive the popular ground swell. Unlike the military intervention of 1898, the political mediation of 1933 did not initially succeed in curbing the popular challenge. The *clases populares* were organized and mobilized, and de Céspedes was too transparently a U.S. pawn. The *machadato* and more than three decades of Plattist politics had exhausted the political class.

On September 4, noncommissioned officers demanding better pay and quicker promotions revolted against the officer corps. The civilian opposition movement turned their insubordination into a military coup. With support from the insurgent officers and the Directorate of University Students, a five-member executive committee formed a government.[10] A week later, Ramón Grau was named president and Antonio Guiteras—a young radical nationalist—minister of government. Without consulting the United States, the Grau-Guiteras administration revoked the Platt Amendment. The United States, the political class, and the Cuban army had been defied.

For four months, the Grau-Guiteras government battled the odds and espoused a nationalist, reformist program. Decrees were passed on minimum wages, an eight-hour workday, utility rates reductions, worker compensation, and collective bargaining. Women were granted the right to vote, and university autonomy conferred. With the 50-percent law, at least half of all employees in all workplaces were required to be Cuban. Bills promoting land reform and the rights of *colonos* against the largely foreign-owned mills were announced. *Cuba para los cubanos* (Cuba for Cubans) reverberated throughout the island. The government sought to establish Cuban control over economic and political life. Nonetheless, after four months, Grau was forced to resign.

Opposition to the nationalist administration covered the political spectrum. Reform did not satisfy the expectations of revolution of the Communist party and the CNOC. The old political class and the new groups that had accepted the Welles mediation feared their banishment from political life, a threat of the Grau-Guiteras government. For the deposed army officers, no compromise was tolerable with the sergeants who had ousted them. Strikes and other working-class actions unnerved U.S. and Cuban capital. Striking workers established soviets in sugar mills that accounted for 25 percent of the harvest. Sugar and nonsugar interests alike joined in opposition to a government seemingly incapable of restoring social peace. The United States, moreover, could not accept Cuban sovereignty. The Roosevelt administration withheld diplomatic recognition and Welles maneuvered to secure an alternative more congenial to U.S. interests. Divisions within the Grau-Guiteras government offered Welles a timely opportunity: Fulgencio Batista provided the wedge to defuse the popular movement.

Having led the September 4 revolt against the army officialdom, Sergeant Batista emerged as the power broker of the Cuban crisis. He sided with the Grau-Guiteras government on two key occasions. In late September, the government violently disbanded a communist-led demonstration in Havana, and confrontations with already striking workers subsequently increased islandwide. In early October, the deposed military officers attempted a coup. Both times Batista reinforced the government, isolating the Guiteras radical faction and increasing the dependence of the moderates on the army. Ambassador Welles took notice and initiated discussions with Batista, who collaborated with Welles in easing Grau from office in January 1934.[11] The anti-Machado revolution and the four-month nationalist government had, nonetheless, transformed Cuba.[12]

Between 1934 and 1940, a new governing consensus was forged. After 1934, when the Roosevelt administration abrogated the Platt Amendment, the United States receded from constant intervention in Cuba. Under Batista and the new officer corps, the army became an arbiter in politics and would no longer be an appendage of incumbent administrations. *Cooperativismo* had tainted the old political parties and even Nationalist Union dissenters failed to gain a permanent place in post-Machado Cuba. Unitl 1940, the anti-Machado faction of the old *políticos* constituted interim civilian governments; Batista and the army, however, wielded the real power in Cuba. When Miguel Mariano Gómez confronted the army on the issue of military control of the educational system, the colonel had enough support in Congress to impeach the president. After 1936, the old political class never again attempted to regain power.

The *Pax Batistiana*, moreover, enlisted forces across the social spectrum. Until 1935, the communist-led CNOC continued to confront capital. In the year following the downfall of the Grau-Guiteras government, more than one hundred strikes, including three general strikes of over 200,000 workers, took place. Batista responded with decisive force and succeeded in retrenching the labor movement. In May 1935, the army assassinated Antonio Guiteras as he was attempting to leave the country. With his death as symbol and the effectiveness of repression, the revolution of 1933 came to an end. The working class, however, was now a factor to take into consideration. During the 1930s, the state continued to pass labor reforms, and Batista allowed the Communist party to rebuild the labor movement. In 1939, the Central Organization of Cuban Trade Unions was founded under communist leadership.[13] Labor was set to play a central role in the emerging consensus.

Batista likewise addressed many reformist demands of the 1920s. In 1937, he announced a three-year social and economic program that included plans for a national bank, support for agricultural diversification, land-tenure guarantees, profit sharing between mill owners and *colonos*, distributing public lands to peasant families, enacting labor legislation, and reforming education and public health.[14] Moreover, in 1940, Batista organized a constitutional convention inclusive of all political sectors and prepared for the restoration of representative democracy. Under the Constitution of 1940, elections were held, and Fulgencio Batista became president of Cuba. Cuban politics was on the threshold of a new logic.

Representative Democracy, the Working Class, and the Emergent Logic

The Batista inauguration embodied both rupture and continuity. The president himself was part of the new generation of political leaders. New social groups—most notably the working class—were incorporated into the mainstream of national politics. The Constitution of 1940 reestablished democracy and reflected a social equilibrium: it legitimized the rights of labor, proscribed latifundia, and assigned the state a central role in the economy while proclaiming the sanctity of private property.

There were still significant continuities with pre-1933 Cuba. Without the Platt Amendment, the United States no longer meddled into every facet of Cuban life. However, sugar quotas bolstered the ties of dependence, and reciprocity continued to reinforce the centrality of an increasingly stagnant sugar sector. Informal consultation supplanted formal intervention. Cuban *políticos* invariably visited the U.S. ambassador for his opinion on a wide range of topics. *Hacendados* and *colonos* routinely traveled to Washington for Department of Agriculture audiences on U.S. sugar quotas. The old political class was marginal but not absent from public life; its members participated in the 1940 constitutional convention and entered electoral alliances with the new political parties. During the late 1930s, Batista reinstated many deposed *machadista* officers. Corruption and graft survived the *machadato* and the upheavals of the 1930s. Public officials continued to view the national treasury as their private domain.

Civil service in public administration was a central tenet in the reformist program. Although the impudence with which *políticos* handled the national treasury injured the public's sense of decency, corruption was not principally a moral dilemma. A state so exclusively mired in a dynamic of enrichment for public officials could not easily implement a program to address national development. As the cases of Mexico and Brazil demonstrated, widespread graft was not incompatible with a developmentalist state. But in Cuba, government corruption did not support economic transformation and, on the contrary, further entrenched the sugar status quo.

With the Sugar Stabilization Institute and the Sugar Coordination Law, the state secured a more equitable distribution of sugar proceeds for smaller mills, *colonos*, and workers. State regulation of the sugar sector, especially the assignment of cane quotas to mills and *colonos*, also created innumerable opportunities for speculation, bribery, and impropriety. Nonsugar interests failed to secure from the state a similar interaction of reform and corruption to promote agricultural diversification and import-substitution industrialization. State actors, moreover, never quite perceived pursuit of a reform program as essential to their interests.

The working class was a potential ally of reform. Agricultural diversification and import-substitution industrialization promised an expansion of employment and the domestic market, benefiting the *clases populares* and the nonsugar sectors of the *clases económicas*. Even so, the reformist alliance faced nearly insurmountable obstacles. Sugar preeminence and trade reciprocity constituted a vicious circle. Present exigencies almost always subdued future prospects. Reinforcement of the status quo took place amid a new balance of social forces, however. In contrast to the early republic, post-1933 Cuba could not relegate the working class to the sidelines.

Although workers had not won the struggles of the 1930s, they were indispensable for establishing a new order. Because the communist-controlled Central Organization of Cuban Trade Unions was included in the consensus around the Constitution of 1940, a dilemma confronted the Cuban state. The Great Depression and World War II bolstered the significance of sugar, which clearly dimmed the outlook for employment opportunities. Reform within the status quo offered the possibility of redistribution but not the likelihood of sustained growth. Yet the working class was organized and militant, and demanded responsiveness. During the 1940s, trade unions pursued a policy of militant reformism. Organized labor provided an important constituency of support for Fulgencio Batista, Ramón Grau, and Carlos Prío, the three presidents elected between 1940 and 1952. The conundrum of employment illustrated well the interaction of militant unions, the *clases económicas*, and the state in a monoculture economy.

The primary objective of the post-1933 union movement was to safeguard employment. Union efforts generally proved effective. Although unemployment and underemployment were never significantly alleviated, job security for those who were employed was virtually guaranteed. Throughout the 1940s, organized labor prevented the modification of a dismissals decree whereby workers could be fired only after cumbersome procedures. During the 1940s, courts decided in favor of labor in three out of five dismissal cases, and the executive regularly decreed wage increases.[15] Militant unions succeeded in maintaining the position of unionized workers and, consequently, made it difficult for capital to improve efficiency.

During World War II, German submarine activity in the Caribbean forced the Cuban government to limit the embarcation of sugar to Havana and Santiago. After the war, workers at other ports won back their previous level of shipments even though it was cheaper to use the larger and more modern facilities in Havana and Santiago. Sugar workers obtained compensation for *superproducción* (surplus production) when mill improvements reduced the number of harvest days. *Hacendados* were obliged to pay workers for the same number of days the mill had operated the previous year. Workers resisted sugar industry attempts to load bulk sugar because it would have reduced employment. Rank-and-file opposition to tobacco industry modernization forced the government to decree compensation for displaced workers after mechanization was approved. One railroad company estimated that 40 percent of its payroll consisted of subsidies and payment for make-work. The *clases económicas* deplored the fact that concessions to labor translated into 70 paid, work-free days a year: 30 vacation days, 11 sick and personal days, 4 holidays, and 27 days from 44-hour work weeks remunerated on the basis of 48 hours.[16]

Many union leaders recognized that their actions were detrimental to long-term development prospects. The memory of pre-1933 conditions and the struggles of the 1930s were, however, fresh in the minds of leaders and the rank and file. Therefore, the CTC took full advantage of its political weight to defend the working-class share of the status quo. Moreover, labor had little confidence that, even if concessions were made, capital would invest in the Cuban economy to create jobs. Domestic investment trends certainly substantiated CTC fears and mistrust. In addition, although capital decried wage increases for their deleterious impact on business, profits and savings increased significantly during the 1940s. In contrast, real wages rose 25 percent and inflation 60 percent between 1941 and 1947.[17]

Collective bargaining rarely settled labor-management disputes. A 1934 decree allowed the state to take over enterprises when labor and capital could not settle their conflicts. Intended as an exceptional recourse, state interventions occurred frequently and often favored workers. The *clases económicas* repeatedly called for legislation and the establishment of labor tribunals to regulate labor-management relations. In effect, their demands ran counter to an executive-dominated political system. Neither the president nor labor had any interest in seeing executive power curtailed. During the 1940s, Batista, Grau, and Prío maximized their rule-making power: the average ratios of congressional legislation to executive decrees were 1:57, 1:70, and 1:26, and labor-related decrees accounted for 13 percent, 18 percent, and 17 percent.[18] Unions preferred presidential decrees and arbitration to legislation and judicial mediation because the executive was more susceptible to immediate political pressure than Congress and the courts.

For the same reasons, the *clases económicas* extolled the virtues of legislative and judicial processes. Cuban capitalists promoted but never obtained the promulgation of a labor code to regulate labor-management relations and minimize executive intromission. They strove to contain the erratic application of Cuban social legislation that they deemed "somewhat quixotic" for a monoculture economy. The *clases económicas* viewed the state as "a major risk for business . . . trampling upon the most basic economic principles."[19] They regarded "interventionism

491

as . . . the professional illness of our public officials," and blamed the state for failure to restore "peace and tranquility . . . to relations between capital and labor after 1933."[20] Sugar and nonsugar interests alike deplored what they perceived to be government partisanship toward labor. *Hacendados* and *colonos* condemned executive-decreed salary increases to sugar workers and were particularly outraged in 1948 when the Grau administration froze wages rather than allow their decline in tandem with sugar prices.[21] Two years earlier the associations of sugarmill owners and cane growers had unsuccessfully sued the government for seizing the *diferencial*—the difference in proceeds from higher sugar prices at the end of the harvest than at its start—and using it for public works and food price subsidies.[22] The industrialists similarly denounced state policies: "Agriculture, industry, and commerce are not charitable activities . . . their reason for being lies in the profit system, and either profits are produced, or all incentive disappears for these economic activities."[23] The *clases económicas* deplored the conduct of Cuban politics and demanded a greater voice in public affairs.[24]

The industrialists, nonetheless, had a more nuanced position toward the working class. In 1945, they sponsored a luncheon with the unions to discuss labor-management relations. CTC general secretary and longtime communist Lázaro Peña was the keynote speaker. *Hacendados, colonos,* other representatives of the *clases económicas,* and the labor minister attended.[25] Peña highlighted four objectives shared by the unions and the industrialists: protection of national industry, creation of a national bank, tax reform, and case-by-case modernization of production.[26] Labor and capital did not agree on the curtailment of state interventions and the liberalization of dismissals, however.[27] The 1945 luncheon underscored ANIC's openness to an alliance with the communist-controlled union movement. Since the early 1940s, the CTC had in fact been calling for cooperation between the industrialists and the working class for the sake of "national unity and salvation."[28] ANIC and CTC disposition notwithstanding, the reform alliance never came together.

The overture of the industrialists aside, the *clases económicas* viewed the union movement with profound hostility. Without success, they encouraged the formation of a second confederation to divide the power of labor. On occasion, however, the industrialists broke rank with the sugar sector. During the 1946 controversy over the *diferencial,* they remained conspicuously silent. Instead of denouncing the state for seizing sugar industry profits, they called for national cooperation in "the economic reconstruction of Cuba," the formulation of a "truly national and healthy economic program," and change in "our colonial economic organization."[29] ANIC, however, could or would not pursue a social pact with organized labor without consensus among the *clases económi-*

cas and generally agreed that union strength and communist control were major obstacles for the mobilization of capital.

The Communist party—renamed the Popular Socialist party (PSP) in 1943—and the CTC were nonetheless central components of the Batista-engineered social peace. And, the PSP and the CTC were—along with the most conservative parties—part of the Batista coalition in the 1940 elections. In 1944, the PSP reached agreement with Grau shortly after the Auténtico party victory. The new president staved off pressures from *auténtico* labor leaders Eusebio Mujal and Francisco Aguirre to oust the communists from the CTC executive committee. Grau could not risk dislodging the communist leadership from the unions without incurring high political costs.

The memory of the 1933–1934 nationalist government stirred popular enthusiasm for the *auténtico* administration. The same memory, however, instigated suspicions among the *clases económicas* and sectors of the armed forces. The administration, moreover, did not have a majority in Congress. A factional struggle over leadership of the union movement would have opened an unnecessary flank. Grau set aside his mistrust of the communists who had not supported his 1933–1934 government and confirmed their hold of the CTC. The PSP had little choice but to accommodate to Grau to maintain its influence in Cuban politics: labor was its principal power base. The 1944 Grau-CTC alliance consolidated the practice of close association between unions and the state that Batista and the communists had initiated during the late 1930s. In 1945, the president awarded the CTC 800,000 pesos for a "Workers' Palace" and banned organization of a second union any place where labor was already organized.

The 1946 election results undermined the rationale for the Grau-PSP alliance. The Auténtico party now had a congressional majority and controlled most provincial and municipal governments. Moreover, two years in power had allowed the party to expand the state bureaucracy and reward supporters with the sinecures of public office. Vanquishing the communist labor leadership would enable the *auténticos* to exercise full control over the union movement and gain favor with the *clases económicas* as well as assuage U.S. concerns over communist influence. The emergent cold war was transforming the international context, and the U.S. government no longer looked unperturbed at PSP domination of Cuban labor. Although a 1945 Office of Strategic Services report emphasized that the communist-controlled CTC did not "attack foreign investments or domestic capital and private property," the State Department expressed apprehension.[30] That same year American Federation of Labor representatives met in Miami with *auténtico* labor leaders to discuss strategies to curb communist influence in the Cuban and Latin American labor movements.[31]

In 1947, the CTC congress provided the occasion for displacing the communists from the labor leadership. The *auténtico* labor commission challenged the credentials of communist delegates who represented local unions recently created by the PSP in order to control the congress and reelect Lázaro Peña to the post of general secretary. Communists rejected these charges and responded with similar allegations against *auténtico* delegates. *Auténtico* labor leaders requested that the Labor Ministry postpone the congress and arbitrate the dispute. Even though willing to seek a compromise with the communists, unaffiliated union leaders like Angel Cofiño and Vicente Rubiera leaned toward the *auténticos*.

Initially, the Grau administration attempted to mediate the dispute and proposed the communists remain in the CTC leadership while turning over majority control to the *auténticos*. In response, Lázaro Peña declared communist support for an independent candidate as general secretary. *Auténtico* labor leaders also acquiesced to a compromise candidate, but refused to accept the communists under any conditions. As a result, communist leaders withdrew from the negotiations, convened their own CTC congress in May 1947, and reelected Peña to CTC stewardship with the support of dissident *auténticos* and some independents. In July, *auténtico* and most independent labor leaders celebrated a separate union congress and elected Angel Cofiño as general secretary. In September, Labor Minister Carlos Prío recognized the July congress and the noncommunist CTC, and the communists lost control of the labor movement. A month later, the PSP withdrew support of the Grau administration.

Purging the communists from the CTC brought an array of consequences—some intended, others quite unanticipated. Leadership struggles and divisions weakened the CTC. The expulsion of the communists did not totally eliminate their influence among rank-and-file workers, especially in sugar, tobacco, and transportation. The *auténtico*-independent coalition, moreover, proved to be short-lived. In 1949, a new CTC congress elected Eusebio Mujal general secretary and consolidated *auténtico* command over the union bureaucracy. Cofiño and other independents formed a separate labor confederation. Removing the communist labor leadership also diminished the weight of the PSP in Cuban politics. Without the unions, the party was rudderless. Militant reformism had run its course.

Expelling the communists did not immediately reduce union militancy and state interventionism, however. Compared to the communists who had led labor struggles for two decades, *auténtico* leaders lacked legitimacy with rank-and-file workers. Consequently, *auténtico* governments continued the policy of intervention and decree rule to strengthen the grass-roots base of their labor leaders. The Prío administration decreed sixty-one interventions—double the number under Grau.[32] Between 1948 and 1952, the wage share of national income climbed steadily from about 60 percent to 69 percent.[33] The 1948 conference on national economic progress noted:

> When the Grau San Martín government decided to break with the communists, the Cuban labor movement was forced to improvise anticommunist leaders without much experience and support among workers. . . . [T]hese leaders have generally sought government concessions that are more radical, costly, and unreasonable than those previously demanded by communist leaders.[34]

More significantly, *auténtico* control of the CTC undermined a crucial contribution the labor movement could have made to a new logic of politics in Cuba. Closely associated with the Batista and Grau administrations, the CTC under communist leadership nonetheless owed its primary party allegiance to the PSP. Never a major party, the PSP did exercise considerable influence in Cuban politics. During the 1940s, about 7 percent of the electorate voted for the communists. Communist deputies and senators distinguished themselves for their discipline, hard work, and honesty. They were only 5 percent of congress, but submitted over 15 percent of all bills.[35] More important, the PSP was not the party in power. The communist-led CTC was more autonomous from the government than the *auténtico* CTC could be. Moreover, Lázaro Peña and other CTC communists had greater weight within the PSP than Eusebio Mujal and other CTC *auténticos* did within their party because the CTC was more important to the PSP than it was to *auténticos*.

The PSP added a new element to Cuban politics: communists were generally honest and collective-oriented. They sought to enhance their power in order to defend what they understood to be working-class interests. Nonetheless, Cuban communists were also realists; in practice, they espoused militant reform, not revolution. From their CTC power base, communists operated in the political mainstream while challenging the predominant logic of corruption. The PSP often switched alliances to retain access to power, and its opportunism undoubtedly blemished its radical credentials. However, PSP effectiveness within the political process strengthened constitutional democracy as the emergent logic of Cuban politics. Communists and the CTC under their control accepted the new consensus while rejecting the tradition of graft and corruption. The PSP-led CTC constituted an essential component for political reform. The *auténtico* CTC did not resist the legacy of the old *políticos:* the new leadership appropriated the union bureaucracy as stepping-stone to public office and fountainhead of personal enrichment. In 1950, an independent leader noted: "The only labor leaders today with integrity and ability are the communists." Similarly, an *auténtico* sugar leader asserted: "We used to get fed up contributing a day's pay to all of the different communist causes, but at least we knew it didn't go into their pockets." Writing in 1952, Charles A. Page noted:

In the fall of 1947, following the pattern of other Latin American labor movements, Cuban labor had become irrevocably split; but the ousting of communist leadership had been . . . accomplished by opportunistic and demagogic chauvinist rivals. . . . Leadership had changed; organization had been broken. The days of the *caciques* have returned.[36]

During the 1940s, *auténtico* administrations failed to consolidate representative democracy and diversify the economy. Their tenure reinforced the old logic of corruption without instituting parallel economic and political reforms. The *auténticos* emerged in the revolution of 1933, and their coming to power initially signaled hope. The former revolutionaries, however, succumbed to the temptation of rapid self-enrichment and sidetracked their erstwhile visions and ideals. Cuba was still a nation of limited economic opportunities, and the next electoral round could drive incumbents from power. The Grau administration expanded the budget to consolidate *auténtico* rule and enhanced discretionary spending. In 1943, for example, the Education Ministry was allotted 15,000 pesos a month to pay the salaries of teachers temporarily without appointment. Three years later the same purpose ostensibly required more than 1 million pesos.[37] When Grau left office in 1948, the Education Ministry handled nearly 2 million pesos a month in discretionary spending.[38] These funds clearly covered more than the salaries of unemployed teachers.

Rampant corruption and widespread disillusionment led to the formation of a new political party. Led by Eduardo Chibás, the Ortodoxo party broke with the *auténticos* in 1947 over the issue of corruption. The new party mobilized a largely middle-class constituency and espoused a program of political and economic reform. Within representative democracy, the *ortodoxos* expressed the sentiments of radical nationalism. Chibás, whose weekly radio broadcasts denounced corruption with incendiary intensity, almost exclusively defined the new party. The slogan *vergüenza contra dinero* (honor against money) more readily identified the *ortodoxos* than programmatic calls for diversification, industrialization, and defense of national sovereignty. Like the PSP and the CTC under communist leadership, the Ortodoxo party defied the old logic. *Personalismo* made an indelible mark on the *ortodoxos*, however. Like most other political parties in Cuba, the Ortodoxo party failed to move beyond a leading personality.

With the support of nearly 46 percent of the electorate, the *auténticos* retained the presidency in 1948.[39] Carlos Prío assumed office without the enthusiasm or the fears that had greeted Grau in 1944. The *auténticos* had passed popular measures such as rent control, protection against land evictions, wage increases, and public works, but expansive misappropriations and the failure to relieve unemployment and inflation had considerably tarnished their popular appeal. Moreover, the *clases económicas* viewed Prío less suspiciously. His record as labor minister under

Grau when the communists were purged from the CTC enhanced his stature in the eyes of Cuban capitalists. Indeed, shortly after his inauguration, Prío addressed the ANIC-Chamber of Commerce conference on national economic progress and assuaged the lingering fears of the *clases económicas:* the new National Bank would decree neither foreign exchange controls nor a currency devaluation. Prío was likewise reassuring that his policy on enterprise interventions and wage increases would be more "reasonable."[40] In an effort to dispel the image of a Cuba inhospitable to U.S. capital, the president sounded the same themes in New York a month later before the Cuban-American Sugar Council.[41]

The increased number of interventions and the growing share of wages in national income during the second *auténtico* administration were somewhat deceptive of the direction that labor-capital relations were in fact taking. The *auténticos* wanted to retain union control and extended concessions to labor to strengthen the new union leadership. But the more frequent recourse to intervention under Prío than under Grau did not necessarily indicate a more favorable climate for labor. The *clases económicas* began to welcome state interventions as a means to convince the government that high labor costs were crippling the economy. Many businesses facing bankruptcy now saw interventions as an opportunity to streamline operations without incurring social and political costs. If the government wanted a company to continue in business, state officials had to impose a compromise on workers.[42] Under Prío, while wages continued to increase, labor-related rule making declined slightly and executive decrees abated significantly. Moreover, prolabor judicial appeals on dismissal cases also dropped. Only one in two cases was now decided in favor of unions.[43]

Prío forged closer ties with Cuban capitalists than Batista and Grau had. His administration formulated policies more favorable to Cuban and foreign capital. Yet the politics of corruption overshadowed state developmental policies. Prío did not take kindly to the World Bank assessment of the Cuban economy. The *Report on Cuba* sharply criticized governmental performance and emphasized the imperative of leadership in taking advantage of prosperity to diversify the economy. Progress was possible only if there was "energetic, resolute, and united action by the Government, by private groups, by individuals, and by the nation as a whole."[44] The *auténtico* government did not quite measure up to the task.

During the late 1940s and early 1950s, increasing violence also weakened the Prío administration and the functioning of representative democracy. Rival *grupos de acción* operated with impunity. During the 1930s, when those responsible for repressing the Machado opposition were not brought to justice, these action groups formed to pursue the *machadistas*. By the 1940s, the *grupos de acción* had long abandoned their political intent and had become gangs defending their turfs and settling scores. Under Prío, the breadth and frequency of their actions grew significantly. Not surprisingly, the early 1950s found a progressively cynical and fearful public. On March 10, 1952, when the military deposed Carlos Prío and Fulgencio Batista once again became president of Cuba, few Cubans bemoaned the demise of the second *auténtico* administration.

The Batista Dictatorship, the Working Class, and Radical Nationalism

The early morning coup was carried out without much resistance. Prío, his family, and closest associates sought asylum in the Mexican embassy and left Cuba shortly thereafter. The *auténticos* were spineless in defending their constitutional claim to power. They had breached their legitimacy while in office, and because of it, representative democracy passed away ingloriously. The coup preempted the elections of June 1952 in which Batista was running a distant third. The *ortodoxo* candidate Roberto Agramonte was favored to win; the *auténtico* Carlos Hevia might have upset him. Neither was particularly inspiring; both were competent and honest. Had those elections been held, reformism might have had another chance. They would also have taken the political career of a young *ortodoxo* lawyer running for Congress down a different path. The candidate, Fidel Castro, was the likely winner in his Havana district.

The idea of a military coup originated with younger officers who wanted to restore order and call new elections. Because during the 1930s Batista had carried out a similar task, they turned to him for leadership. Moreover, after becoming president in 1940, the general established civilian control over the armed forces. In 1952, Batista sidetracked the expectations of the younger officers. Contrary to the 1930s and 1940s, Batista did not have much popular support. To remain in office, he had to rely on the armed forces, especially the officers who had joined him in the sergeants' revolt on September 4, 1933, and who were now anxious to reap the bounties of power after twelve years of civilian rule. The March coup renewed the political preeminence of the military in Cuban politics.[45]

The *clases económicas* generally welcomed the overthrow of the Prío administration. General Batista promised to restore order, and his record during the 1930s lent additional credibility to his words. Initially, however, most citizens reacted with indifference. Some students protested and the union movement called a general strike without success. In early 1953, University of Havana professor Rafael García Bárcena conspired with young *ortodoxos*, university students, and sectors of the officer corps to engineer a coup against Batista. They were discovered and imprisoned. Opposition plans also flourished among mainstream *ortodoxos* and *auténticos*. Without shared visions and concordant tactics, the older *polítocos* quarreled among themselves. Former presidents Grau and Prío vied for personal control of the Auténtico party. The issues of a united front with *auténticos* and the efficacy of electoral politics to undermine Batista divided *ortodoxos*. In mid-1953, some *auténtico* and *ortodoxo* leaders met in Montreal and signed a "unity" pact supporting negotiations and elections. The Montreal Pact proved to be inconsequential, serving only to underscore *auténtico* and *ortodoxo* disarray: they were incapable of effective leadership, common purposes, and united action. Restoring the Constitution of 1940 became the rallying cry of the slowly mounting Batista opposition.[46]

Like other *ortodoxo* youths, Fidel Castro was outraged at the news of the military coup. He presented a legal brief in the Court of Appeals in Havana demanding imprisonment for Batista and his collaborators for violating the constitution. Not unexpectedly, the court disavowed the request. Shortly thereafter Castro rejected negotiations with Batista as a means to bring an end to the dictatorship and became a leading proponent of armed insurrection. In 1953, the centenary of José Martí's birth, 165 young Cubans heeded Castro's call. For months, preparations proceeded with great secrecy and spartan dedication. Most participants learned the details of the plan two days before the action: they were to seize Moncada Barracks in Santiago de Cuba, distribute arms to the population, and spark a national insurrection.

At dawn on July 26, after most *santiagueros* had reveled in a night of carnival, the attack against Moncada Barracks took place. It was a resounding fiasco. Dozens of youths were captured, tortured, and killed, the rest imprisoned. The nation was horrified by governmental repression and moved by the daring, if reckless, action of the young Cubans. Fidel Castro especially captured the popular imagination. When brought to trial, he defended himself with integrity, compassion, and dignity, and sketched a political program of nationalist reform. Although he and his surviving comrades were sent to jail, Castro bowed only to the judgment of history: "Condemn me, it does not matter. History will absolve me!"[47]

In 1954, Batista called elections. Grau entered the contest but withdrew when it became evident that the outcome was never in question: Batista would remain in office. Though unopposed, the general used the election to claim legitimacy for his rule. With a new sense of stability, Batista made some concessions. He convened Congress and allowed most political parties to resume their activities. In May 1955, the government declared a general amnesty and released all political prisoners. Among those freed was Fidel Castro, who immediately resumed his opposition activities, exploring the possibilities of peaceful struggle. Less than two months later, however, he went into exile reaffirming his belief in insurrection against the dictatorship. In August, the July 26th Movement issued a manifesto to the people of Cuba:

The Cuban Revolution does not compromise with groups or persons of any sort. . . . [I]t will never regard the state as the booty of a triumphant group. . . . [W]e assume before history responsibility for our actions. And in making our declaration of faith in a happier world for the Cuban people, we think like Martí that a sincere man does not seek where his advantage lies but where his duty is, and that the only practical man is the one whose present dream will be the law of tomorrow.[48]

The July 26th Movement was now separate and distinct from the Ortodoxo party, and Fidel Castro was its central figure.

After the 1954 elections, the government took steps to spur the transformation of the economy and issued the National Program for Economic Action. State interventions to settle labor disputes virtually stopped. Courts supported management against labor on dismissals appeals in three out of four cases.[49] Domestic and foreign capital found a more auspicious climate, and business confidence increased. Gustavo Gutiérrez, the old *machadista* reformer, prominent economist, and author of the new program earnestly hoped for the permanence of reform: "Rulers pass on, the Republic continues."[50]

The *batistato*, however, more decidedly accentuated the dynamics of corruption. Between 1952 and 1956, government revenues totaled more than 1.3 billion pesos; Grau and Prío had collected 1.9 billion in eight years.[51] During the 1950s, public works expenditures alone added up to more than 1 billion pesos. Less than 50 percent covered actual costs; the rest, commissions and profit margins.[52] The new development banks granted Batista supporters generous loans and declined modest requests from nonpartisans. Malfeasance and graft were even more widespread in the regulation of the sugar industry. *Hacendados* and *colonos* who condemned the new levels of corruption suffered reduced quotas and fewer business opportunities. The Sugar Stabilization Institute became nothing more than a forum to reward Batista supporters and punish opponents.[53] The military regime subordinated the incipient development infrastructure to the logic of corruption.

After failing to mobilize the rank and file in a strike against the coup, the union leadership reached an accommodation with Batista and thus continued the CTC tradition of harmonizing with the government. Workers had long developed an "opportunistic tolerance" for the shifting ideologies of their leaders.[54] In supporting Batista, Eusebio Mujal breached his *auténtico* affiliation. During the 1950s, the union movement generally complied with state efforts to create a more favorable investment climate. Most significantly, the CTC agreed to a long-standing demand of the *clases económicas:* the modification of dismissal procedures. In exchange, the government granted unions the right to compulsory payroll deduction of dues, increasing the amount of cash at their disposal. Wage demands were "generally not immoderate," and the CTC often overlooked the curtailment of vacations and other benefits.[55]

During the 1950s, although the wage share of national income declined from 69 percent to 64 percent, it was still higher than in 1948.[56] The establishment of new labor relations more conducive to economic transformation proceeded slowly.

Nonetheless, the *mujalista* CTC lacked the militance and relative independence of the communist-controlled union movement during the 1940s. Accommodationism supplanted militant reformism. The CTC leadership concurred in effect, if not always in rhetoric, with the consensus of the *clases económicas* that labor conditions partially obstructed economic transformation. Yet unemployment and underemployment persisted, and capital-intensive development did not promise relief. CTC accommodation to the *batistato* further divided organized labor. High-ranking union leaders opposed the revision of the dismissals policy.[57] In 1955, sugar workers went on strike to demand their share of the *diferencial.* The CTC leadership ordered a return to work before reaching an agreement with the government. Local union leaders and rank-and-file workers, however, continued on strike until a compromise was attained. Similarly, Havana bank workers' leaders defied the *mujalista* bureaucracy in their actions for salary increases and other benefits. In both instances, the CTC purged dissident union leaders.[58] In general, the CTC reinforced its stronghold in Havana while provincial unions loosened their ties to the national union bureaucracy. During the 1950s, national trends pointed to widening disparities between Havana and the rest of the country, and a similar polarization characterized the union movement.

CTC support for the government notwithstanding, many workers joined the ranks of the anti-Batista movement. Some strike leaders like Conrado Bécquer, José María de la Aguilera, and Reynol González subsequently joined the July 26th Movement. The *diferencial* strike, moreover, demonstrated the potential of mobilizing widespread support for worker actions against Batista. Small businesses and industries in dozens of provincial towns and cities closed their doors in sympathy with the demands of sugar workers. The *diferencial* represented additional purchasing power that the private sector did not want to lose.[59] In 1957, workers paralyzed Santiago in protest against the assassination of Frank País, second in command of the July 26th Movement. The PSP organized labor committees that rivaled official unions in nearly three hundred enterprises.[60] In 1958, however, the July 26th Movement called for a general strike that succeeded in Santiago and other provincial cities but failed in Havana, where the CTC and the government were clearly in charge. Subsequently, communist and July 26th Movement labor leaders formed a united front that would be instrumental in the general strike of January 1, 1959. Like the *mujalista* CTC, however, the emergent parallel union movement was not a militant organization akin to the communist-led CTC during the 1940s.

In 1955, a series of developments marked the anti-Batista movement. *Auténticos, ortodoxos,* and other *políticos* regrouped and seemed to
be better coordinated. University students elected new leadership and
expressed renewed discontent. Toward the end of the year, independence war veteran Cosme de la Torriente formed the Friends of the
Republic Society and called for a civic dialogue and a new round of
elections. Except for the July 26th Movement, every other opposition
sector participated. Although Batista accepted the invitation, he would
not concede to elections before their scheduled date of 1958. His intransigence bolstered those who argued that armed struggle was the only
way to challenge his rule.

After the civic dialogue failed, violence from within and from
without confronted the government. Colonel Ramón Barquín and other
professional military officers sympathetic to the *ortodoxos* conspired unsuccessfully to depose the general. Hard-line officers also failed to ease
Batista from power. In May 1956, a group of *auténticos* attacked Goicuría
Barracks in Matanzas. On December 2, the *Granma* carrying Fidel Castro
and his followers landed in the southern coast of Oriente. The political
stability that Batista had thought he had achieved in 1955 would be
short-lived.

Against extraordinary odds, the *Granma* expeditionaries reached the
Sierra Maestra and survived the first encounters with the army. In urban
areas, the July 26th Movement stepped up actions and, along with other
urban-based groups, bore the brunt of widening repression. Throughout
1957 opposition mounted. In March, under the leadership of José Antonio Echevarría, the Revolutionary Student Directorate (DRE) attacked
the Presidential Palace with the objective of assassinating Batista. The
students came perilously close to success. Carlos Prío financed an
aborted landing of his men in Oriente to establish a site of armed resistance independent from the Sierra Maestra. DRE members who had
disagreed with the Batista assassination attempt opened a minor guerrilla front in the Escambray mountains in central Cuba. Manifestos ema-
nated from all opposition quarters and pacts were forged to combat the
dictatorship. In the Sierra Maestra, Fidel Castro met Raúl Chibás,
brother of the founder of the Ortodoxo party, and Felipe Pazos, former
president of the National Bank. They issued a joint communiqué advocating a civilian provisional government, restoration of civil liberties,
social and economic reforms, and abstention by other nations from intervention in the Cuban crisis. In September, a naval mutiny in Cienfuegos unsettled the armed forces from within. Governing Cuba became
increasingly difficult for Batista.

During 1958, the general's problems intensified. The United States
imposed an arms embargo against his government but at the same time
failed to encourage the moderate opposition.[61] Under the auspices of the
Catholic church, moderates called for a coalition government and a new

dialogue without much success. Like Gerardo Machado two decades earlier, Batista became more intransigent as momentum gathered against his rule. Nonetheless, there were moments when defusing the opposition seemed a plausible prospect. In April, the general strike failed, in part because the government had effectively contained the July 26th Movement.

The military could not follow up with similar achievements against the Rebel Army, however. A summer offensive against the *rebeldes* did not stall their ascendance. In July, opposition forces signed a new agreement that recognized armed insurrection and a general strike as the primary means to combat the dictatorship. Raúl Castro consolidated a second guerrilla front in Oriente. Ernesto Guevara and Camilo Cienfuegos marched toward the Escambray mountains in central Cuba. PSP-supported guerrillas had earlier established themselves in the Escambray. The DRE guerrillas, an independent July 26th outfit, and a motley band of semibandit bands also operated in Las Villas.

During the 1930s, the Cuban armed forces had gained expertise in combating urban insurrection. After 1956, Batista skillfully marshaled that experience. But rural Cuba, especially Oriente, where land squatters were more common and landlords more successful in evicting peasants, was another matter.[62] The Rebel Army first gained the upper hand merely by surviving and later by resisting the regular army. The Sierra Maestra, however, provided the *rebeldes* with a skewed view of Cuba: like Havana, the eastern mountains were not representative of Cuban society. Moreover, their quick victory would lead them to emphasize military over civilian skills. Military prowess, however, did not ultimately defeat Batista.

⊱ A climate of collapse enveloped Cuban society. By 1958, multiple crises—social, political, economic—besieged Cuba, and it was no small irony that the maelstrom trapped Fulgencio Batista. During the 1930s, Batista had forged a new governing consensus overcoming popular upheavals and elite dissension. During the 1950s, the general proved incapable of mustering his past experience and enlisting a variety of social and political sectors in preserving what would soon become the old Cuba. Twenty years had passed and quite a few opportunities had been missed. Nonetheless, by refusing to relinquish power, Batista ultimately undermined the moderate opposition and braced the radical nationalism of Fidel Castro, the Rebel Army, and the July 26th Movement. ⊱

In spite of the arms embargo, the United States vacillated in condemning the government and encouraging the opposition, especially because Castro—a man the U.S. government never quite trusted—was becoming the dominant figure.[63] Moreover, with the Platt Amendment long abrogated, the Eisenhower administration was not inclined to mediate between government and opposition as Roosevelt had done in 1933. In December 1958, the mission of William D. Pawley was only a

half-hearted attempt to replicate the Welles mediation. By then, even a
more concerted effort would probably have come to naught. The *fidelistas*, who had never supported negotiations with Batista, were on the
verge of victory and would not have welcomed U.S. auspices.

The anti-Batista movement undoubtedly mobilized diverse social
and political forces. Moderate oppositionists twice sought negotiations
and elections. In October 1956, Fidel Castro and José Antonio Echevarría signed a unity pact of mutual recognition of their organizations in
the struggle against Batista. The urban and *sierra* factions of the July
26th Movement coexisted tensely over matters of tactics and organization. Frank País was working toward a more formal organization, with
shared leadership between the Rebel Army and the urban July 26th
Movement, when he was assassinated. País considered the general
strike—not guerrilla warfare—the catalyst for a national insurrection.
The July 1957 Sierra Maestra declaration was followed by a November
Miami manifesto reiterating a united opposition program of constitutional restoration and socioeconomic reforms. The Miami pronounce-
ment, however, equivocated on denouncing foreign intervention in Cuban internal affairs and proposed the integration of the Rebel Army with
the regular army after the downfall of Batista. Claiming improper Sierra
Maestra and July 26th Movement representation, Fidel Castro condemned the new manifesto.

By 1958, the Rebel Army and the July 26th Movement were undis-
putedly at the helm of the Batista opposition. A midsummer pact among
opposition forces recognized armed insurrection as the principal means
of struggle. The PSP also supported the Sierra Maestra guerrillas and
formed labor committees with the July 26th Movement. Although liberals, radicals, and communists coexisted loosely within the opposition
movement, events had proven the advocates of armed insurrection right.
Without them, the struggle against the dictatorship would not have been
where it was. By then, too, design, chance, and talent had made Fidel
Castro the uncontested leader of the national insurrection.

It was far from inevitable that Fidel Castro and the Sierra Maestra
guerrillas would command the heights of the anti-Batista struggle. *Auténticos, ortodoxos,* and the old *políticos,* for example, might have succeeded in forging a united front against Batista. The general might have
consented to free and honest elections and ushered in a provisional
government in late 1955 when Cosme de la Torriente led the civic dialogue movement or early in 1958 when the Catholic church revived it.
José Antonio Echevarría might have survived the March 13, 1957, Presidential Palace attack, and the Revolutionary Student Directorate might
have then exacted a more equitable relationship with the *fidelistas*. Had
Frank País not been assassinated, the urban July 26th Movement might
have exercised greater direction of the struggle than the Sierra Maestra
guerrillas.

Within the context of the Cuban 1950s, the July 26th Movement was not a reformist movement. In his 1953 self-defense, Fidel Castro outlined the program the *moncadistas* would have implemented had they been successful: restoration of the Constitution of 1940, agrarian reform, profit-sharing in industry, greater share of sugar industry profits for *colonos*, and confiscation of misappropriated wealth. He defined the Cuban people as the unemployed, rural workers, industrial laborers, small farmers, teachers and professors, small merchants, and young professionals. *Hacendados*, large landowners, and large commercial and industrial interests were conspicuously absent. The proscription of latifundia and the promotion of full employment expressed objectives already mandated in the Constitution of 1940.[64] Other July 26th Movement documents highlighted agrarian reform, industrialization, and the extension of education and health care.[65] The July 26th Movement economic thesis favored a program of active state intervention and protection of domestic capital over foreign investments.[66] Indeed, the substance of these proposals constituted the kernel of reform in other Latin American countries. But not in the Cuba of the 1950s.

During the 1890s and the 1930s, Cuban elites with help from the United States had derailed challenges from below. Longer-term political stability had, however, proven much more elusive. The Platt Amendment frustrated the first republican effort, and the record of the *auténtico* administrations had debilitated representative democracy. Moreover, the social and political dynamics of the 1950s did little to bolster the forces that could have sustained reform in the wake of Batista. Ending the dictatorship and restoring the constitution were the expressed opposition objectives, but Fidel Castro and the July 26th Movement insisted more clearly on the tactics of insurrection than on the specifics of the new Cuba. Although their programmatic statements were often vague and contradictory, they were clear and explicit on the character of the July 26th Movement as an organization that repudiated the past and aimed to renovate Cuban politics.[67]

The rallying cry of the Batista opposition—the restoration of the Constitution of 1940—was not a call for the status quo before the military coup. The constitution symbolized the ideals of democracy, social justice, and honest government that representative democracy during the 1940s and dictatorship during the 1950s had traversed. The July 26th Movement was unequivocal in rejecting foreign intervention and demanding that the Rebel Army be the sole guarantor of the new Cuba. The *fidelistas* called for change in a society where political and economic failures had considerably weakened the recourse to reform, and they used radical means to secure power. That armed struggle provided the final blow against the dictatorship further undermined the possibilities of reform after the revolution came to power.

Fidel Castro and the July 26th Movement were intransigent in their

summons for national regeneration. José Martí was their mentor; concluding the nineteenth-century quest for *Cuba libre* their purpose. By
January 1, 1959, the structures that could have restrained their intransigence were barely in place, clearly fragile, and often discredited. The
overthrow of Batista brought radical nationalism to power, and Fidel
Castro and the *rebeldes* owed allegiance only to *el pueblo cubano*. Over
the next two years struggles between the *clases populares* and the *clases
económicas* as well as confrontations with the United States radicalized
the revolution. Remembering 1898 and 1933, the revolutionaries of
1959 refused to compromise; instead, they mobilized the working class
and the *clases populares* and forged a new consensus based on national
sovereignty and social justice. ⟩

CHAPTER 3

Revolution and Radical Nationalism, 1959–1961

This time Cuba is fortunate: the revolution will truly come to power. It will not be as in 1895 when the Americans intervened at the last minute and appropriated our country . . . It will not be as in 1933 when the people believed the revolution was in the making and Batista . . . usurped power . . . It will not be as in 1944 when the masses were exuberant in the belief that they had at last come to power but thieves came to power instead. No thieves, no traitors, no interventionists! This time the revolution is for real!

Fidel Castro
Santiago de Cuba
January 1, 1959

On New Year's Eve, Fulgencio Batista fled Cuba. The Cuban people jubilantly welcomed the revolution. "This is a decisive moment in our history: tyranny has been defeated. Our happiness is immense, but we have much yet to do," Fidel Castro told the nation on January 8.[1] The new government, however, did not have a clear blueprint for the future. Fidel, the Rebel Army, and the July 26th Movement repudiated the past and were committed to national regeneration. What was being repudiated and what the process of renewal would entail was another matter. The pursuit of social justice and national sovereignty took precedence over the restoration of the Constitution of 1940. By 1961, the Cuban economy was no longer capitalist and new forms of politics were emerging. After January 1, a new coalition of elites and social forces emerged to consolidate the revolution.

With the revolutionary government committed to their interests, the *clases populares* acquired a new sense of empowerment. Their support galvanized the radicalization, which, in turn, alarmed the *clases económicas* and the United States. Neither, however, proved capable of containing the revolutionary onslaught; six decades of mediated sovereignty and growing political crisis had undermined them, and *el pueblo cubano* was willing to consider the making of a new Cuba without them. Thus, the leadership of Fidel Castro, the mobilization of the *clases populares*, and the defense of the nation against the United States were the catalysts of revolutionary politics.

Reformism, the Clases Económicas, and the Revolution

On January 4, 1959, Fidel Castro named Judge Manuel Urrutia president of the revolutionary government. The new cabinet assembled the best of liberal Cuba: lawyers, judges, economists, *ortodoxos*, veterans of the 1930s, participants in the Rafael García Bárcena movement and the civic dialogue of the 1950s, social activists—Cubans who spoke their nationalism reasonably and moderately. Their program was one of economic growth, more equitable distribution of wealth, honest government, due process of law, and the pursuit of national interests. They were similar to the reformers of the 1920s who had initially supported Gerardo Machado in the hopes of making Cuba wealthy and Cuban. Unlike them, however, these liberal Cubans did not constitute a movement: a general strike inaugurated their government and the Rebel Army backed it. "Power is not the fruit of politics, but the fruit of sacrifices by hundreds and thousands of our comrades. Our commitment is only to the people and the Cuban nation," Fidel Castro noted on January 1 in Santiago de Cuba.[2] Indeed, politics—the *politiquería* (political chicanery) that had perverted elections and undermined constitutional rule—was not a source of legitimacy for the revolutionary government. Those who had died in the struggle against Batista, *el pueblo*, and the quest for *Cuba libre* were.

The Rebel Army and the July 26th Movement were not solely responsible for overthrowing Batista, but Fidel Castro and the *rebeldes* commanded the opposition movement after the summer of 1958. Although the July 26th Movement had benefited from the association with the moderates, the latter had joined the *fidelistas* after attempts to oust Batista on their terms had failed. Many wealthy Cubans, whose contributions had totaled 5 to 10 million pesos, had also supported the insurrection.[3] Although the promises to respect capitalism and restore constitutional government had partially assuaged their misgivings, they still had lingering doubts. The *rebeldes* had been quick to burn cane fields and extort supplies from large landowners. Known for his autocratic tendencies, Fidel Castro was reluctant to share the mantle of leadership. Moreover, unswerving commitment to armed struggle and resolute repudiation of the past were the only clear positions of the *fidelistas*. And yet, on January 1, 1959, Fidel Castro, the Rebel Army, and the July 26th Movement were incontestably the liberators of Cuba, and virtually all Cubans supported them. Victory was not theirs alone, but the unfolding of the anti-Batista struggle had rendered them—not others—indispensable.

Liberal Cuba had no claim on power independent from Fidel Castro.

How the *rebeldes* would compose the new Cuba was a different matter. Their only statement on the economy was far from radical: it merely advocated an activist state on behalf of economic development. Authors Felipe Pazos and Regino Boti underscored the importance of economic growth and social justice. Pazos and Boti—subsequently National Bank president and economy minister—berated long-standing beliefs about the Cuban economy. They rejected the immutability of monoculture, the terms of Cuba-U.S. relations, the impossibility of industrialization, the scarcity of domestic capital, and the primacy of foreign investment. They called for agrarian reform, sugar industry modernization, import-substitution industrialization, and investments of state and domestic capital. Their objectives were to foster full employment and economic growth, and to redistribute national income.[4]

The economic thesis of the July 26th Movement read like a program for reform, yet it was different from the reformism that had gained consensus among the *clases económicas* during the 1950s. The July 26th Movement program was not antilabor; Pazos and Boti dismissed the notion that labor legislation subverted economic development and defended the compatibility of high wages with economic growth. They faulted the lack of economic planning for the slow movement toward industrialization, diversification, and job creation. Indeed, they supported a far greater degree of state intervention than the prevailing consensus. Whereas the Batista program had been most interested in foreign investment, Boti and Pazos called for the primacy of Cuban public and private capital. The July 26th Movement thesis was committed to national control of the Cuban economy.

Had negotiations brought an end to the Batista regime, the consensus of the 1950s might have led to "tropical dependent development." Armed struggle, however, had toppled the dictatorship. The July 26th Movement program espoused reformism, but the social forces that were its natural constituency—the middle class, the industrialists, and nonsugar agricultural interests, the more privileged sectors of the working class—lacked independent political leadership. During the 1940s, the *auténticos* had compromised their claim to represent a Cuba of greater sovereignty, economic progress, and social justice. The coup of 1952 had prevented the Ortodoxo party from governing and the possibility that radical nationalism might have unfolded under the Constitution of 1940. In 1959, nationalist reformers had no leader other than Fidel Castro, no movement but the July 26th, no army but the Rebel Army. They lacked the resources to direct the popular ground swell that welcomed the revolution toward a program of reform.

Nonetheless, the *clases económicas* also joined in celebrating the revolution. *Hacendados, colonos,* cattle ranchers, tobacco and rice growers, industrialists, commercial establishments, individual businesses, and domestic and foreign corporations extended their congratulations to the new government in daily media advertisements. Indeed, the *rebeldes*, their red-and-black flag, and Fidel so completely impacted the national imagination that by March, the registry of industrial property was swamped with petitions to patent products bearing the revolutionary trademarks.[5] The symbols of revolution were deemed good for business.

Not all sectors of the *clases económicas* applauded the revolution with equal enthusiasm nor welcomed its program wholesale. *Hacendados* and cattle ranchers were especially wary of the promulgation of agrarian reform, and importers opposed protectionism. The industrialists embraced the policies of industrialization, tax and agrarian reforms, and tariff revisions but cautioned against spiraling wages. While supporting agrarian reform, *colonos* balked at raising the minimum wages of rural workers. Rice growers appreciated higher domestic quotas but requested a one-year postponement of wage raises.[6] The revolution intensified the historic divisions among the upper classes.[7] Their principal organizations were either riven by dissent, like those of the *hacendados* and the *colonos*, or beset by traditional weaknesses, like those of the industrialists. The *clases económicas* proved as ineffectual against the revolution as they had been defending national interests before 1959: their earlier failures had decidedly undermined their rampart against radical nationalism.

The initial program of the revolutionary government was not exceptionally radical in form. Its central edict—agrarian reform—was based on article 90 of the Constitution of 1940, which proscribed latifundios and further foreign land ownership.[8] The reform allowed landholdings of up to 1,000 acres, as well as special allotments to 3,333 acres if economically justifiable. Compensation would be in 20-year bonds at 4.5 percent interest. About 10 percent of all farms were affected. Foreign and sugar-mill ownership of land was subsequently prohibited.[9] The new law also built upon the Sugar Coordination Law of 1937, securing the rights of *colonos* against *hacendados*. Moreover, its land-to-the tiller motto clearly favored small *colonos* and peasants over agricultural workers. Indeed, the agrarian reform virtually ignored the 500,000 Cubans in the rural labor force.[10] Its provisions for the establishment of cooperatives rather than the redistribution of land and a strong role for the state were, however, more suspect. The National Institute for Agrarian Reform (INRA) absorbed the old sugar, rice, and coffee institutes and became a powerful new state institution.

Other measures were likewise within the realm of reform. Progressive tax policies favored Cuban over foreign investments, nonsugar over sugar sectors, small over large businesses, the provinces over Havana. State regulatory powers were exercised to benefit small producers: smaller sugar mills and small rice growers were assigned larger quotas.

Rents were reduced 30 to 50 percent, but landlords earning less than 150 pesos in rent income a month were excluded.[11] Although controversial, stiff taxes on imports and foreign exchange controls were no more radical than the Economic Commission for Latin America recommendations of Raúl Prebisch. Indeed, in eight months, wage increases, new jobs, and various other reforms generated an expansion of 200 million pesos in domestic purchasing power.[12] New investment inquiries were up tenfold, license applications for small businesses increased 400 percent. As national industries expanded their production of consumer goods, U.S. exports to Cuba declined 35 percent.[13]

From the start, the revolution distinguished between the industrialists and the other sectors of the *clases económicas.* Calls for national unity in defense of popular interests included them.[14] Denunciations against those who had supported the Platt Amendment and trade reciprocity and forsaken national control of the economy abounded. But so did praise for the Cubans of wealth who had invested in industry and agricultural diversification. "Their interests coincide with those of the nation," noted an editorial in *Revolución,* the July 26th Movement newspaper.[15] At a large industrial enterprise in Havana, Raúl Castro commended the "comrade owners" for meeting worker demands and making donations to the National Institute for Agrarian Reform.[16] The peasantry, the working class, and the progressive bourgeoisie were identified as the three pillars of the revolution.[17] Landowners and importers—not workers—were the enemies of the industrialists.[18] The government applauded the initiative of workers and owners at Santiago harbor to create a committee in defense of national sovereignty.[19] Indeed, workers and managers often expressed joint support of the revolution.[20]

The National Association of Cuban Industrialists welcomed inclusion in the ranks of the revolution. While balking at wage increases, ANIC supported industrialization and submitted its program to the revolutionary government.[21] The industrialists requested a meeting with the CTC to coordinate efforts in support of the government and discuss disagreements over labor-management relations.[22] The industrialists stressed that the monthly salary in industry averaged 2.5 times the minimum wage. Industry, moreover, had no dead season.[23] The CTC never agreed to the meeting.[24] In May, when agrarian reform alarmed other sectors of the *clases económicas,* the president of ANIC noted:

> We avoided involvement in yesterday's corrupt politics, but now we have to cooperate with today's good politics because in four months the revolutionary government has done more for Cuba than was done in all our previous years of republican life. . . . We are under the obligation to contribute to the government's achievements. There is absolutely no excuse not to do so.[25]

On January 1, 1960, ANIC congratulated the revolutionary government on its first anniversary and praised the industrialization program, administrative honesty, domestic market expansion, and import-export regulations. The industrialists also volunteered to wage an international campaign to improve the credit image of Cuba.[26]

The effects of the revolutionary program were, however, profoundly radical. More than Batista fell on January 1, 1959. As after the overthrow of Machado, Cuban society confronted the threat of a radical transformation. Only now there was no Sumner Wells, no regular army, no full spectrum of organizations vying for space in the new Cuba—no means, in short, to contain radical nationalism. The revolution of 1933 sealed the fate of early republican politics; the revolution of 1959 eventuated from the failure to govern Cuba beyond the Platt Amendment and trade reciprocity. The passage of agrarian reform galvanized a popular ground swell that engulfed the nation and found support in Fidel and the Rebel Army. When the new leadership rejected immediate elections as a brake to the revolution, most Cubans were not concerned. The past had demonstrated that *politiquería* easily scuttled the integrity of electoral processes and the interests of the *clases populares*. The revolutionary government saw a new popular *conciencia* based on the attainment of social justice and national sovereignty as the best guarantee of democracy.[27]

Moreover, many nationalist reformers supported the revolution and went along with its early measures. They agreed to postpone elections; the first cabinet of liberal citizens centralized enormous and unchecked powers in the cabinet. They were generally silent when revolutionary justice violated due process in the trials of Batista collaborators. Few objected when Fidel Castro ordered the retrial of a group of air force pilots who had been acquitted of war crimes and when the new trial condemned them.[28] Nationalist reformers supported the program of agrarian reform, industrialization, and employment expansion because it was also their program, and they were bereft of the leadership, the organizations, the institutions, indeed, the historical resources to transform Cuba on their terms.

Opposition to the revolution stirred in foreign and domestic quarters. The United States sharply condemned the revolutionary trials, especially the retrial of the Batista pilots. The agrarian reform alienated the sugar sector and U.S. interests. *Hacendados* and cattle ranchers launched a media campaign against the reform.[29] As the new power contours emerged, the *clases económicas* began to disinvest.[30] They could not accept that "the radical transformation of social structures which has taken place in Cuba . . . means that their opinions are no more valued than those of ordinary citizens."[31] In February 1959, Fidel made an affirmation that did little to soothe the fears that the revolution awakened even in the industrialists:

A truth needs to be stated and that is simply that workers are right. It must be stated so that no one in our country has absolutely any doubts about it. Workers, . . . the unemployed have paid the harshest consequences of past failures. . . . Our economy has developed within a system of private enterprise, . . . consequently, workers are not responsible for our desperate straits. Only our immoral governments and the wealthy who opted for unproductive investments are.[32]

Never before in Cuban history had a government so unabashedly favored the *clases populares*. The reorganization of the Central Organization of Cuban Trade Unions became a central focus of revolutionary politics.

The Working Class and the Revolutionary Government

Like Batista, Eusebio Mujal and most of his close associates fled Cuba; exile was their only alternative to revolutionary justice. The recently formed July 26th Movement–PSP labor committees immediately took over the CTC. July 26th Movement labor leaders controlled the CTC executive; the PSP assumed many rank-and-file positions. Many *mujalista* labor leaders also retained control of local unions. July 26th Movement labor leaders were generally younger and less experienced than either the *mujalistas* or the communists. They were also less numerous and could not completely fill the vacuum in the CTC after the flight of many *mujalistas*. Thus, communists and *mujalistas* stepped in where the July 26th Movement could not.

In early 1959, a torrent of demands for better wages and improved working conditions underscored the extent of CTC concessions to the emerging consensus of the 1950s that a more compliant labor movement was necessary for economic development. Workers demanded 20 percent across-the-board increases, an immediate renegotiation of labor contracts, and the reinstatement of workers fired for political reasons.[33] In Santiago, workers called for an equalization of wages with Havana, and soon labor assemblies in the provinces seconded the call.[34] The National Federation of Sugar Workers (FNTA) called upon *hacendados* to pay workers the *diferencial* of 1958 and for *superproducción*—up to 50 million pesos in wages lost due to mechanization and innovations after 1953.[35] The FNTA also demanded the institution of four work shifts in the mills to alleviate unemployment.[36] Numerous union assemblies demanded vacation payments that management had illegally retained.[37] Strikes were relatively common, and strike threats even more so.[38] At the end of January, CTC General Secretary David Salvador observed: "We prefer harmonious solutions to all conflicts, but where owners close all possibilities for such resolutions, we will go on strike however many times it is necessary. . . . [W]e will not make anti-revolutionary concessions to employers who conspire against the revolution."[39]

The *clases económicas* resisted the renewal of working-class militancy. *Hacendados* and *colonos* slowed down the 1959 sugar harvest.[40] The industrialists opposed immediate salary increases and renegotiation of labor contracts.[41] Some union leaders and militant workers were fired; management sometimes refused to meet union representatives.[42] Many enterprises denied union activists the time needed to attend to labor matters; management had regularly granted labor leaders time off for the same purpose during the 1950s.[43] *Hacendados* challenged the legality of unions.[44] Some *colonos* closed food stores, limited credit, and cut back jobs in protest over wage increases.[45] Others publicized spurious labor contracts with wages set below the daily minimum of 3.14 pesos.[46] Enterprise lockouts were relatively common.[47] Many capitalists purposively delayed or otherwise obstructed Labor Ministry mediations with unions to settle new contracts.[48] At least once, management encouraged workers to strike to subvert these mediations.[49] One enterprise established a company union.[50] The *clases económicas* experienced the unraveling with unsettling swiftness of the more favorable ambience in labor-management relations that had been emerging during the 1950s.

The revolutionary government wasted no time in seizing the initiative. Hampering the economy in any way was considered to be unpatriotic. The Labor Ministry expected full cooperation between workers and employers and often mediated conflicts to preempt strikes and lockouts.[51] In early 1959, the ministry conducted more than 5,000 mediations, supported wage increases on a case-by-case basis rather than across the board as the unions demanded, and proclaimed "equidistance" between the interests of labor and those of capital.[52] Nonetheless, the Labor Ministry was not a neutral arbiter. Without granting all demands, mediations generally settled conflicts in favor of workers. In ten months, the revolutionary government decreed 66 million pesos in wage increases to sugar workers and 20 million pesos to workers in other sectors.[53] In 1956–1958, wage increases had averaged 4.2 percent; the annual rise in 1959 was 14.3 percent.[54] The cabinet ordered the reinstatement with back pay of workers fired for political reasons during the 1950s and the disbursement of illegally retained vacation payments, and awarded sugar workers retroactive *superproducción* wages.[55] Moreover, the government revived the old practice of interventions to prevent plant closures, settle labor conflicts, and enforce labor legislation.[56] Between 1934 and 1952, the state had resorted to intervention 101 times,[57] and after 1952, Batista had virtually stopped the long-standing practice. In eighteen months, the revolutionary government intervened more than 200 times, in most instances at the request of workers.[58]

Indeed, the *clases económicas* had reason to doubt the Labor Ministry's claim to equidistance. The tempo of daily events and the tenor of revolutionary statements could not but shake their confidence. In an address to Shell Oil Company workers who had threatened to strike if wage increases and other demands were not satisfied, Fidel Castro, while arguing that strikes were not appropriate under the revolution, asserted:

> Workers are the principal creators of wealth, not the capitalist in his comfortable Wall Street office. . . . The revolution is yours and for you. We are going to wage large battles, not small battles. Larger, more useful and beneficial than those we have discussed tonight. Now the people are in power. We are a single entity. People and power are a single entity.[59]

From the outset the Rebel Army supported the *clases populares*. Provincial commanders often intervened in the conflicts of sugar workers with *hacendados* and *colonos* and settled the differences in favor of FNTA. The agrarian reform fueled the upsurge in rural Cuba and the Rebel Army, INRA, and FNTA emerged as a powerful triumvirate on behalf of rural workers. "For the first time, the army will not use its weapons against the people," Fidel Castro noted.[60] His brother Raúl affirmed: "The Rebel Army is a political army whose purpose is to defend the interests of the people."[61] These were not uncertain words, and the actions that followed them were even less equivocal. For the first time in Cuban history, the *clases económicas* were without any army. Nonetheless, more than the Rebel Army was needed to direct the popular ground swell. With the collapse of the *mujalista* hierarchy, CTC reorganization became imperative. In the process, an amalgam of tensions, issues, and contradictions between the July 26th Movement and the PSP quickly surfaced.

The PSP had not played a major role in the anti-Batista struggle. The party had condemned the Moncada Barracks attack as "putschist," "adventurist," and "against the interests of the people." In the mid-1950s, the party had fomented the formation of parallel unions that turned out to be important in the emergent anti*mujalista* movement. In the summer of 1958, Carlos Rafael Rodríguez had met with Fidel in the Sierra Maestra. A small group of PSP-supported guerrillas in Las Villas had joined the Rebel Army columns of Ernesto Guevara and Camilo Cienfuegos in their westward drive. In the fall, the July 26th Movement and the PSP had constituted united labor committees. Nonetheless, the communists—not unlike the moderate opposition—had endorsed the armed rebellion when other avenues of struggle against Batista had all but disappeared.

The party had had an uncanny record of survival. In 1939, when the CTC was founded, communists had hailed Batista in spite of his earlier repression of the popular movement. During the 1940s, they had supported the Batista and Grau administrations. After the 1947 CTC purge, PSP relevance to national life had diminished considerably. Even had the 1952 elections taken place and the *ortodoxos* won with PSP endorsement, the party was unlikely to regain its pre-1947 prominence. Without the labor movement, the communists had lost their platform in Cuban politics. Indeed, the triumphant revolution whose origins communists had dismissed as "putschist" and "adventurist" thrust their party into the mainstream. The PSP had an organization, able leaders, experienced cadres, and international allies. None had been decisive in the anti-Batista movement. All, however, would prove to be crucial in forming a new governing coalition. The *mujalista* collapse created the opportunity for communist labor leaders to reclaim their previous space in Cuban society. The CTC vacuum also afforded July 26th Movement union leaders the chance to establish their realm within the revolution.[62] For twenty years, the CTC had represented a constituency whose interests the state had to address. The revolutionary government was no different.

The July 26th Movement–PSP labor front was short-lived. At the end of January, the July 26th expelled the PSP from the CTC executive committee.[63] The action might have been occasioned by the anticommunism of the July 26th Movement labor leaders. Like the purges of remaining *mujalista* leaders, the communist ouster from the CTC leadership, however, more likely was caused by the July 26th's determination to gain exclusive control of the trade unions.[64] Revolutionary Student Directorate labor cadres also joined the anti-PSP struggle. After January, July 26th Movement labor leaders called for union elections and the celebration of a CTC national congress.[65]

Local elections were held in April and May. The preelection period was often marked by turmoil. Rank-and-file demands for better wages and working conditions prompted Labor Ministry mediations to prevent strikes. The revolution insisted upon cooperation with capital for the sake of the nation. In February, Fidel Castro unequivocally stated:

> We have to defend the revolution with more than just a simple demand. The revolution is the demand of today and that of the future, the salaries of today and those of the future, the welfare of today and that of the future. . . . Strikes are formidable weapons, but we cannot use them now.[66]

In April, CTC General Secretary David Salvador congratulated workers, union leaders, and management at a Havana enterprise for resolving their differences peacefully, and criticized trade unionists who

> adopt extremist attitudes like declaring strikes and work slowdowns or occupying factories. These are measures which do not contribute to the best defense of workers' interests. . . . [W]e have a revolutionary government which guarantees the demands of Cuban workers. . . . No measure which encumbers the revolutionary government . . . is good for the working class. . . . [T]o the extent that the revolution consolidates its power to that same extent workers will be strong and we will be guaranteeing . . . permanently and with a sense of history our present interests, our future interests, our interests always.[67]

That the revolutionary government favored workers in most mediations bolstered July 26th Movement labor leadership: more than 90 percent of all rank-and-file leaders elected belonged to the movement.[68]

The PSP often posited a leftist challenge to the CTC. PSP General Secretary Blas Roca observed: "When strikes are necessary and just, they are not harmful, but helpful to the revolution."[69] Communists criticized the government for treating capital and labor on the same terms.[70] Before the union elections, PSP labor cadres took over the leadership of many local unions in railroads, sugar, transportation, shoes, graphic arts, and docks, among others.[71] In some instances, communists delayed the elections; in others, they refused to hand over the unions to the new leadership.[72] PSP labor leaders accused many July 26th Movement labor leaders of election irregularities, antidemocratic practices, and *mujalista* connections.[73] Communists denounced the CTC for remaining in the AFL-CIO–sponsored Inter-American Labor Organization.[74] Although the PSP mustered considerable experience and talent in the attempt to reclaim its old mainstay, its success was limited.

The July 26th Movement parried the PSP offensive in the labor movement. For months after the union elections, *Revolución* published scathing indictments of communist history: the compromise of the 1933 revolution, the collaboration with Batista during the late 1930s and the early 1940s, the reformist transformation of the labor movement, the participation in *politiquería* and back-room deals, the inability to forge a strategy against Batista, the opportunism at most turns in national politics, and the antinational dependence on a foreign power.[75] With considerable reason, the July 26th Movement dismissed the PSP call for unity as a strategy to retain relevance after a resounding electoral defeat. Moreover, the past rendered the communists less than trustworthy allies. Like the revolution, the CTC belonged to those who had led the struggle against Batista. Workers, moreover, had turned down the PSP bid for union leadership.

After Fidel Castro proclaimed "humanism"—neither capitalism nor socialism—as the ideology of the revolution, the July 26th Movement labor leadership followed suit. The most anticommunist faction formed the Front of Humanist Workers (FOH) to defend July 26th Movement control of the labor movement. Twenty of the thirty-three federations associated with the CTC subscribed the FOH. The CTC leadership strongly supported agrarian reform, denounced incipient terrorism against the revolutionary government, and underscored the social functions of property.[76] The July 26th Movement–dominated CTC invoked a six-month no-strike pledge.[77] When workers and their local unions grumbled in some sectors against it, the CTC censured them and clamped down on dissent.[78] In September, Fidel Castro stated that demands for salary increases were no longer legitimate: the national economy, unemployment, and the welfare of *los humildes* (the poor) were more important.[79]

The conflict between the July 26th Movement and the PSP reached a climax at the CTC congress in November. No more than 10 percent of the 3,200 congress delegates were communists.[80] National events, however, had taken a hectic turn in the weeks preceding the congress. In October, a new labor minister was sworn into office. Former defense minister and guerrilla commander Augusto Martínez Sánchez claimed the right to regulate labor-management relations in whatever ways necessary.[81] With the entire country on the lookout for Camilo Cienfuegos, the popular *comandante* of the Rebel Army whose plane disappeared over the skies of Camagüey, Martínez Sánchez postponed delegate elections to the upcoming congress. CTC leaders expressed concern that they had not been consulted about the postponement.[82] As counterrevolutionary activities increased, national defense became the overriding imperative. The Rebel Army faced internal dissent in Camagüey when Huber Matos resigned his post of provincial commander to protest growing communist influence. Cienfuegos had gone to Camagüey to arrest Matos. Likewise, other personnel changes pointed to new directions. Raúl Castro became defense minister, and Ernesto Guevara president of the National Bank; both were known to be sympathetic to the communists. The CTC congress opened in a sharply more polarized national ambience than that of April–May when the local union elections were held.

The congress called on the working class to support the revolutionary government and the agrarian reform, contribute to industrialization and the elimination of unemployment, defend national sovereignty, and promote racial integration.[83] While endorsing improvement in living and working conditions, the agenda did not include demands for wage increases and shorter workdays. Fidel Castro emphasized that defending the nation and consolidating the revolution were now more important: "The destinies of *la patria* and of the revolution are in the hands of the working class."[84] Similarly, he implied that cooperation between the July 26th Movement and the PSP was crucial to the struggle against the enemies of the revolution. Nonetheless, an overwhelming majority of the delegates voted down the PSP candidates to the executive committee. The issue, however, was not settled.

Some delegates continued to support a unity slate. Many insisted that the PSP call for unity was a communist maneuver to gain a foothold in the revolution. Discussions heated up and no consensus seemed possible. Then, Fidel Castro returned to the congress and chastised the delegates for their near-riotous behavior. What would have happened had the delegates been armed? The proceedings were undermining the morale of the working class. Fidel said he also had a right to speak for the July 26th Movement, and he was speaking on behalf of unity among CTC leaders. Like the nationalist reformers, anticommunist labor leaders lacked a base of popular support separate from that of Fidel Castro and the revolution. Moreover, they had earlier eschewed an alliance with the industrialists that might have bolstered their position within the CTC. The delegates agreed to an executive committee that excluded the communists as well as the most prominent members of the Front of Humanist Workers.[85] The congress also created a committee to eradicate the remnants of *mujalismo*.[86]

By the end of 1959, the revolutionary government had restrained the trade unions. Well before the elimination of private property, defending the revolution and contributing to economic development were the primary goals of the labor movement. A united CTC was critical in the mounting confrontation with the *clases económicas* and the United States. The commitment of workers to national goals rather than to "economistic" demands was likewise vital. The communists were seasoned labor leaders, and their links with the Soviet Union enhanced the value of the PSP as an ally in the radicalization of the revolution. Indeed, the call for CTC unity served the purposes of the revolutionary government well.

After the congress, the Labor Ministry used the anti-*mujalista* commissions to purge uncooperative union leaders. Undoubtedly, some *mujalistas* still occupied posts in local unions throughout Cuba. As the conflict with the communists intensified, many of them embraced the July 26th Movement. Early in 1960, CTC General Secretary David Salvador criticized the anti*mujalista* campaign: because there were relatively few *mujalistas* in local unions, the commissions were primarily expelling independent union leaders.[87] In fact, the anti*mujalista* crusade removed about 50 percent of the labor leaders elected in April and May of 1959, and veteran PSP labor cadres took over many local unions. In May, Salvador was no longer at the helm of the CTC. In 1960, Fidel Castro and the revolution enjoyed such widespread support that the *mujalistas*, independent local leaders, and Salvador could have easily been removed from office by holding new elections: the slate of candidates sanctioned by Fidel and the government would have easily won. Nonetheless, the choice was to purge them.

Rank-and-file workers in construction, restaurants, tobacco, transportation, and utilities resisted the curtailment of union autonomy and their "economistic" demands.[88] Utility workers were especially reluctant to forgo their privileges and briefly confronted the revolutionary government.[89] Although also manifesting some opposition, sugar workers more readily consented to the new cooperation between labor and the government.[90] In June 1960, the CTC executive committee proposed a wage freeze, suggested salary reductions if necessary for national development, and discarded references to different sectors within the working class. Henceforth, the "Cuban proletariat" would be the only valid expression.[91] Fidel Castro succinctly stated the need for a new attitude among the working class: "We have to teach workers to think as a class and not as a sector. . . . We have to teach them to think not only of those who are working, but also of those workers who do not have a job."[92]

The 1959 labor congress resembled the congress of 1947, which had purged the communists and paved the way for the then-*auténtico* Mujal to gain control of the trade unions. Both congresses witnessed fierce elite struggles over CTC stewardship. Both also experienced substantial intervention from the Labor Ministry. In both congresses, the state needed a more docile labor organization to advance national development. Both were also instances of the state's exercising greater command of the labor movement to consolidate national power. Although in 1947 and 1959 sectors of the rank and file resisted the changes, most did not. In 1947, Cuban workers did not rally behind Lázaro Peña against Angel Cofiño and Eusebio Mujal. In 1959–1960, they similarly failed to support the Front of Humanist Workers at the CTC congress and even David Salvador when he was purged.

In 1959–1960, however, Cuba was radically different than in 1947. The revolutionary government could legitimately claim to be the first in Cuban history to uphold the interests of the *clases populares*. No other government had done so much to improve popular living standards in so short a time: indeed, *el pueblo* and the government appeared to be one. By the end of 1960, moreover, the state commanded the major means of production. Cuba no longer had a capitalist economy, and the *clases económicas* were now history. Working-class support was everywhere evident as workers safeguarded their work centers against sabotage and readied to defend the nation against U.S. aggression. How trade unions would function under socialism and how the new state would represent the interests of the working class and the rest of the *clases populares* were yet to be settled. In 1959–1960, Cuba was also radically different than in 1947 because the revolution was eliminating autonomous political action. Unions—never too independent before 1959—would now be submitted *in toto* to the logic of revolutionary politics.

Revolutionary Politics and the Clases Populares

Fidel Castro and the new Cuban leadership did not see their legitimacy dependent on restoring the Constitution of 1940 and holding elections; redeeming Cuba from a past of indignities and improving the welfare of the *clases populares* sanctioned the revolutionary government. Before 1959, the *clases económicas* had accepted a state of affairs that had surrendered national sovereignty and contravened the interests of a majority of Cubans. The revolution had subverted the old order and was creating a Cuba of equality, full employment, agrarian reform, public health, universal education, real democracy. *Cuba para los cubanos* once again reverberated throughout the island, and *el pueblo cubano* had unbridled hopes for the future.

Those calling for immediate elections had the most to lose by the redistributive policies of the new government. When Batista had announced the National Program for Economic Action in 1955, the United States had not responded by pressing democratic reforms on him. Likewise, most of the *clases económicas* had then been more interested in maintaining order than in calling elections. No sooner had the revolutionary government come to power than sectors of the *clases económicas* and the United States started demanding immediate elections. Their sudden appreciation of democracy too clearly belied their primary concern that the new Cuba would not attend to their interests. Why hasten elections when the revolution was rendering Cuba so much more democratic than ever? *"Politiquería* is as odious as tyranny," Fidel Castro asserted in May 1959.[93] Before holding elections, the new government would promote employment, expand health care, extend education, create a new popular *conciencia* about politics. Until then, elections would only put a brake on radical transformation.

Nonetheless, new institutions were needed in order to govern; the question was what kind and for what purposes. If elections were to be held, political parties had to be organized. The postponement of elections, however, supposed continuing the revolution outside the bounds of normal institutional processes. In 1959, some July 26th Movement leaders sought to organize their movement into a political party.[94] Fidel Castro never encouraged those steps; even though it was then inconceivable that the movement would or could have challenged him, he probably resisted the creation of an organization that in the future could have contested him. Still, the fact of the social revolution itself also encouraged alternate forms of politics. Party building amid growing confrontations with the *clases económicas* and the United States would have further exacerbated the conflict with the communists, and the PSP was crucial to the emerging coalition in support of the radicalization. Then, too, without a new *conciencia* about politics, the premature transformation of the July 26th Movement into a party would have likely reinforced the political culture of *politiquería*. Some movement offices— *casas del 26*—were already becoming centers for the dispensation of favors and the transaction of deals reminiscent of the old *políticos*.[95] In the spring of 1959, turning the July 26th Movement into a political party would have been obligatory had elections been on the agenda. Elections, however, had been postponed. Moreover, the support of the *clases populares* and most other Cubans for the revolutionary government was so resounding that advocating elections seemed a formality.

Fidel Castro himself was the most powerful political resource of the revolution. He had an exceptional ability to interpret and address the reality of Cuba, as well as an extraordinary capacity to impress upon his followers the magnitude of their mission. Fidel had proven his talents in the opposition to Batista and did so once again in 1959–1961 as weather vane to the revolution. The first few months of 1959 crafted what the newspaper *Revolución* called a "Fidel-*pueblo* binomial," which became the crux of revolutionary politics. One of the editorials noted that Fidel used a clear and new language: he explained, the people understood.[96] The revolution was now the Sierra Maestra, *el pueblo* the Rebel Army, the United States and Cubans without national dignity the enemy. Just as the *clases económicas* witnessed the almost instant withering of their domination in early 1959, so did the *clases populares* sense the birth of a new power—inchoate, amorphous, unstructured, but nonetheless real. And Fidel infused them with the purpose to conclude the task so often frustrated: the constitution of Cuba as a nation.[97] Most Cubans would then have hardly disagreed with the editorial stand of *Revolución*.

The dynamics of revolution brought forth the new institutional order. The Rebel Army, INRA, and the FNTA supervised the transformation of rural Cuba. The Labor Ministry, the CTC, and the popular militias directed the mobilization in the cities. In June 1959, as *colonos* closed stores and denied credit to rural workers, the government opened the first string of *tiendas del pueblo* (people's stores).[98] In October 1959, when a group of citizens cleaned and painted the Havana waterfront in preparation for a convention of travel agents, volunteer work was born.[99] In February 1960, as relations with the *clases económicas* deteriorated, the central planning board JUCEPLAN was formed.[100] In March 1960, the Association of Rebel Youths (AJR) enlisted the considerable energies and enthusiasm of Cuban young people.[101] In September 1960, after the Organization of American States implicitly rebuked Cuba and foreign intervention seemed imminent, Fidel Castro convened the Committees for the Defense of the Revolution (CDR).[102] Founded in August 1960, the Federation of Cuban Women was also an organization created amid the social revolution.

In his first address to the nation on January 1, Fidel Castro mentioned the need to overcome discrimination against women, especially in the labor force.[103] Almost immediately, the Labor Ministry began to enforce labor legislation regarding women more strictly. New regulations were also enacted with respect to the right of pregnant women to their jobs and the duty of employers to secure a safe environment for women. Enterprises with more than fifty women workers had the obligation to provide separate rest rooms and lounges.[104] In April, Revolutionary Feminine Unity, a July 26th Movement affiliate, convoked a congress with the purpose of organizing women to defend the revolution and their specific interests.[105] After the congress, many women became more active in support of the agrarian reform, the distribution of school lunches, the mobilization of peasants to Havana for the celebration of July 26, 1959, and the organization of the AJR.[106] Like their male counterparts, women workers were especially militant. The Labor Ministry reported "significant increases" in the number of complaints concerning violations of the rights of working women in the first five months of 1959 in comparison with the same period in 1958.[107] In October 1959, July 26th Movement women held another congress to organize a Christmas toy campaign and continue their undertakings on behalf of young people and the agrarian reform.[108] In November 1959, Vilma Espín presided over a delegation of Cuban women to a congress of Latin American women in Chile, and the group formed the core of the future Federation of Cuban Women.[109]

Mobilizing the support of women did not generate internecine struggles, as it had for the CTC. The feminist movement of the 1920s and 1930s had dissipated after women had been granted the suffrage in 1934 and the Constitution of 1940 had guaranteed legal equality. Subsequently, there was no feminist organization analogous to the CTC. After 1959, Cuban women joined the revolution without bringing with them old organizational ties. Various organizations emerged without much apparent conflict over leadership or ideology. The July 26th Movement–PSP debate had little bearing among politically active women. They seemed completely focused on issues of social justice and national defense. Like most men, most Cuban women were not communists and, indeed, were anticommunists on January 1, 1959. Nonetheless, they strongly supported the revolution and followed a similar course of radicalization. By the end of 1960, the call for unity that in 1959 had been dismissed by some CTC leaders as PSP opportunism resounded with veracity and urgency to most Cubans. Without unity, the revolution would not survive. In August 1960, various organizations came together in the Federation of Cuban Women. Among these were all July 26th Movement women's affiliates, various fronts of humanist women, PSP-associated groups, and a Catholic organization *Con la cruz y con la patria* (With the Cross and the Homeland).[110]

In May 1960, before a million Cubans in Havana, Fidel Castro officially announced that the government would not hold elections. His audience shouted that the people had already voted, and they had voted for Fidel.[111] The revolution polarized Cuba and disallowed neutrality. *Con Cuba o contra Cuba* was the battle cry, and Fidel was the embodiment of Cuba. Cuba was not Guatemala, Fidel no Jacobo Arbenz, the Rebel Army indivisible.[112] Anticommunists tended to be more concerned over the conflict with the PSP than about the struggle against the *clases económicas* and the United States.[113] Though opportunists, the communists did not own the national wealth nor were they promoting aggression against *la patria*. Moreover, the anticommunist alarm was only partially sounded against the relative prominence of PSP cadres in the revolution. During the 1940s, communists had controlled the unions and held numerous public offices. Although the United States and the *clases económicas* disapproved, PSP presence then had not precipitated a crisis as it was now doing.[114] More profoundly, the controversy over communism masked the repudiation of radical change. A humanist ideology against capitalism and communism so eloquently espoused in the spring–summer of 1959 was a casualty of domestic and foreign confrontation. Had Cuba not been ninety miles from the United States, the revolution might have found those elusive middle grounds. That nearness and the historic intimacy it had imposed between the two countries had, indeed, contoured the radical nationalism that was now rendering the revolution so intransigent.

Centralization of power quickly became a concomitant of the revolution. The spectre of 1933 hovered ominously in the *rebelde* memory of history. Then, the revolutionaries fragmented, the United States mediated, and the Grau-Guiteras government vacillated in uniting supporters and could not withstand opposition. Divisions and lack of resolve had turned the revolution into a *jugarreta* (a bad play).[115] There was also Guatemala during 1950–1954: a reformist government had encountered fierce domestic and U.S. opposition and had failed to take steps to defend itself. Not as prominent but no less relevant was Bolivia during 1952–1956, when accommodation to local capitalists and foreign interests compromised a social revolution. So that the Cuban revolution would not be compromised, decisive and effective central authority was mandatory. Fidel Castro never hesitated to exercise it.

When members of the Revolutionary Student Directorate retained their weapons after January 1, Fidel ably disarmed them. Why was there a need for weapons independent of the Rebel Army? Was there not a revolutionary government in power to guarantee order and popular interests? Was the victorious revolution going to tolerate the emergence of action groups like those of the 1940s?[116] The first cabinet of liberal citizens soon discovered that the real loci of power in Cuba were Fidel Castro and the Rebel Army. Throughout 1959, different circumstances and various reasons occasioned the demise of the liberals within the revolutionary government. Fidel became prime minister upon the resignation of José Miró Cardona; Osvaldo Dorticós assumed the presidency when Manuel Urrutia objected to mounting radicalization; Augusto Martínez Sánchez replaced the social democrat Manuel Fernández in the Labor Ministry; Ernesto Guevara took over the National Bank from Felipe Pazos. One after the other nationalist reformers fell as the revolution surged without patience for middle grounds or tolerance for dissent.

The organizations and associations of the old Cuba also fell before the revolutionary onslaught. Most perished ingloriously as their history and the force of the revolution prevented them from staking their claim in the new Cuba. Some reorganized and became integrated into the revolution: small *colonos* in the Cane Growers Association, for example, formed the National Association of Small Peasants (ANAP). Like the CTC, the Federation of University Students (FEU) yielded to revolution-

ary direction after considerable conflict. After the October 1960 nationalization of industry and commerce, the revolution entered a new stage.
The PSP, the DRE, even some fractions of old *auténticos* and *ortodoxos*
joined the July 26th Movement in a loose coalition that in 1961 became
the Integrated Revolutionary Organizations (ORI). Cuban politics was
acquiring the contours of a one-party system and, in the process, the
revolutionary government was establishing a new authority. Because
the new politics was contrary to the long-standing basis of Cuba-U.S.
relations, the U.S. government was quite taken aback.

The United States welcomed the *fidelista* victory against Batista with
some reluctance.[117] The composition of the new government was reassuring, but its character was almost immediately disquieting. No real
power was invested in the moderate cabinet that was the only potential
arena of U.S. influence. President Urrutia was indecisive and usually
waited for the opinion of Fidel Castro before making decisions. That "a
fully stable, organized, and responsible government" was not readily
emerging after the downfall of Batista was a central U.S. concern.[118]
Moreover, U.S. official and media condemnation of the trials of Batista
collaborators gave Fidel Castro cause to launch the first anti-American
campaign, further fueling U.S. mistrust. Like their Cuban counterparts,
U.S.-owned enterprises resisted wage demands and union reorganization. Their resistance, however, inflamed nationalism and placed rank-
and-file demands in a highly charged context. In March, when the revolutionary government took over the U.S.-owned Cuban Telephone
Company, the intervention became a symbol of nationalism and popular
defiance.

In April, Fidel Castro traveled to the United States on an unofficial
visit. Had the U.S. government offered aid, he might have accepted it;
had he requested it, the United States might have granted it. Neither side
took the initiative. Shortly thereafter, the State Department and the CIA
decided that it was "impossible to carry on friendly relations with the
Castro government" and started to "devise means to help bring about
his overthrow and replacement by a government friendly to the United
States."[119] In May, the agrarian reform elicited serious U.S. misgivings
concerning the stability of Cuban sugar supplies, the immediate compensation of U.S. holdings, and the long-term ambience for U.S. firms in
Cuba. Although the United States recognized the right of the Cuban
government to enact the reform, the expressed misgivings underscored
the past tenor of Cuba-U.S. relations, which the revolutionary government was adamantly intent on changing. That the first air raids against
cane fields and the initial acts of sabotage in urban Cuba were conducted
with tacit U.S. support no doubt contributed to the nationalist intransigence of the revolutionaries.

The year 1960 opened with a crescendo of mutual confrontations. In
February, Anastas I. Mikoyan visited Havana to inaugurate a Soviet
trade exhibition. A five-year trade agreement for an annual delivery of 1

million tons of sugar and the extension of $100 million in credits to purchase industrial equipment followed the Mikoyan visit. In March, when the ship *La Coubre,* loaded with weapons that Cuba had acquired in France, exploded in Havana harbor, the revolutionary government blamed the CIA. In the memorial to the victims, Fidel displaced *libertad o muerte* as the rallying cry of the revolution and proclaimed the more radical stance of *patria o muerte.* The revolution was now struggling to safeguard *la patria,* without which freedom was unthinkable. In June, when Texaco, Shell, and Standard Oil refused to refine Soviet crude petroleum, the government confiscated their holdings. In August, after a series of mounting economic measures against Cuba culminated in the elimination of the sugar quota, U.S. properties were nationalized. In September, the United States prompted the Organization of American States to rebuke the Cuban government. The revolution reached out for new allies against aggression; as ties with the Soviet Union and Eastern Europe expanded, so did the importance of the PSP to the new governing coalition.

Before departing from office in January 1961, Dwight Eisenhower severed diplomatic relations with Cuba. On April 17, when John F. Kennedy dispatched an invasion force of Cuban exiles to overthrow Fidel Castro, his administration expected to succeed, just as the U.S.-inspired coup against Jacobo Arbenz in Guatemala had in 1954. Unlike Guatemala, Cuba was experiencing a social revolution with profound historical roots and extraordinary popular support. The United States, moreover, never provided the air support necessary to establish a beachhead. Two days later, the revolutionary forces repelled the invasion and made prisoners of most of the invaders. When Cuba stood up against the United States and won, on April 19, 1961, the Platt Amendment *de facto* expired. On April 16, Fidel had declared the socialist character of the Cuban Revolution.

After 1959, socialism became an alternative in the heat of the social revolution. As confrontations between workers and capitalists prompted state intervention, the structures of capitalism considerably weakened. Workers and the rest of the *clases populares* did not explicitly clamor for socialism. Their mobilization in the sociohistorical context of Cuba in 1959 did, however, allow the revolutionary leadership the socialist alternative. Herein lay the internal dynamics of radicalization as the politics of revolution centered on Fidel Castro and popular mobilizations. Once these dynamics became dominant, survival was the central question. How could radical revolution endure in Cuba when dependence on the United States was so pervasive? Part of the answer was the centralization of power and the elimination of independent political activity. The cold-war world provided the other part: an alliance with the Soviet Union.

The interaction of domestic and foreign factors in the radicalization of the Cuban Revolution was complex. No single component determined the course of affirming Cuban independence from the United States and pursuing social justice for the *clases populares*. By 1959, it was perhaps too late for the United States to redefine its relations with Cuba and for the revolutionary government to seek an acceptable compromise: history conditioned their responses, and each had a diametrically different reading of it. Fidel Castro and the *rebeldes* were committed to *Cuba libre*, and the United States had never encountered such determination from a Cuban government. Consolidating a nationalist revolution led Cuba to socialism, an alliance with the Soviet Union, and permanent hostility from the United States.

After 1959, a social revolution momentously unfolded in Cuba. "The 26th of July is the revolution of the *humildes*, for the *humildes*, and by the *humildes*," Fidel Castro affirmed in March of 1960.[120] Commitment to remedying the social injustices of the past radicalized the Cuban Revolution. Intransigence polarized the *clases económicas* and antagonized the United States. Never before in Cuban history had those in power been so defiant and so insistent. "Cuba is not a simple geographical reference," proclaimed a *Revolución* editorial.[121] *Con Cuba o contra Cuba* meant *con Fidel o contra Fidel*.[122] Indeed, in 1959–1961, radicalization, polarization, and centralization consolidated the revolution around Fidel Castro. The new Cuba embarked on a quest of historic proportions with tangible rewards in rising standards of living for the *clases populares*, as well as with more ethereal bequests of dignity and honor. In June 1960, Fidel exclaimed:

> The revolution shows that ideals are more powerful than gold! If gold were more powerful than ideals, those large foreign interests would have swept us off the map; if gold had more power than ideals, our *patria* would be lost because our enemies have plenty of gold to buy *conciencia* and yet all our enemies' gold is not enough to buy the *conciencia* of a revolutionary. . . . Workers, peasants, Cubans of dignity have conquered their revolutionary *conciencia*. . . . They will not trade their revolution, their *patria* for gold.[123]

Over the next three decades, the Cuban government would grapple with the consequences of victory. Confronting the challenges of governance was now the question. National affirmation against the United States would be the overriding consideration, and, therefore, survival would subsume all other concerns. For the sake of *la patria*, ironhanded unity behind Fidel Castro would be enforced. Thus, politics acquired a sense of military discipline contrary to political diversity and independent organizations. Centralization of power and curtailment of autonomy accompanied the politics of survival. Formal democracy—the processes of contestation and rotation—had decidedly limited vistas. The new politics did, however, allow the state to direct the national economy and partially relieve the sense of insufficiency that had permeated the old Cuba. Indeed, the revolution had much yet to do.

CHAPTER 4

Revolution and Inclusive Development

What does Cuba expect to have in 1980? A per capita income of $3,000, more than the United States has now. And if you don't believe us, that's all right too: we're here to compete. Leave us alone, let us develop, and then we can meet again in twenty years, to see if the siren song came from revolutionary Cuba or from some other source.

Ernesto Guevara
Punta del Este, Uruguay
August 8, 1961

The victory of January 1 . . . turned the people into sole owners of their wealth. . . . Revolutionary laws imparted them with material benefits, but above all the Revolution bestowed upon them, for the first time in their history, the conquest of their full dignity, *conciencia* of their power and of their immense and inexhaustible energy.

Granma *editorial*
December 2, 1986

The revolution infused Cuban society with a new logic: the interests of the *clases populares* were now at the center of national development. The new government also disavowed the subordination of Cuba to the United States. Cubans were more equal, the nation more sovereign. However, success in satisfying basic needs raised popular expectations for more comfortable living standards, and daily life in socialist Cuba disappointed them. Relative equality amid austerity was the most prominent socioeconomic achievement. In addition, the Cuban economy still depended on sugar exports and a single market to earn foreign exchange. Thus, socialism did not sufficiently develop the economy to secure national independence. Moreover, with the collapse of the Soviet Union, the international conditions that had supported the consolidation and survival of the Cuban government disappeared.

The revolution also aimed to foster a new *conciencia*. Without a transformation of popular consciousness, Cuban leaders contended that socialism was no better than capitalism. They argued that the exclusive pursuit of individual, material well-being contravened the commitment to equality and the imperative of national unity against the United States. Although the revolution unearthed a valuable resource in the will, energy, and passion of the Cuban people, the challenge was turning them into an economic force in daily life and work. At the beginning of the 1990s, the Cuban government continued to insist on the viability of socialism and the imperative of *conciencia* in maintaining the social achievements of revolution and independence from the United States.

Development Strategies and Economic Performance

The revolution generated an unbounded optimism about the future: at last, Cuba would realize its potential. Diversification was the key to economic growth, employment expansion, and income redistribution. With proper leadership and relentless determination, national independence and social justice would be more than lofty ideals. The revolution mobilized the nation in ways unforeseen—though in intent not unlike those hoped for—in the old Cuba. The revolutionary government turned to strategies of economic diversification and sugar industry modernization that the reformers had long advocated and other sectors of the *clases económicas* had more recently subscribed.

That sugar alone no longer offered realistic prospects for development was, indeed, part of the emerging consensus of the 1950s. No significant progress had yet been registered, but the momentum away from monoculture was mounting. In 1956, the U.S. Commerce Department had noted the need for "a high degree of cooperation" among the various social sectors and "sound and aggressive leadership" from the government for Cuba to forge and implement an effective policy of diversification.[1] The revolution impounded the social cooperation and the state directives then in the making and instituted new directions. After 1959, state policies reversed the trends in favor of Havana and against the *clases populares*, prevented the restructuring of Cuban ties with the United States, and preempted the renewal of dependent capitalism. Nonetheless, the *rebeldes* pursued policies that, in important ways, did not differ from the frustrated reform program.

During the 1950s, there had been little appreciation in Cuba, or elsewhere in the Third World, of the obstacles that stood in the way of development. The Economic Commission for Latin America raised expectations that turned out to be unfounded; the world economy rapidly moved in directions different from those it had anticipated. Globalization took precedence over domestic market expansion in nations like Mexico and Brazil. Caribbean and Central American countries partially modernized their agricultural sectors while experiencing some industri-

alization for export and limited domestic consumption. Latin American
links to the international economy were transformed without apprecia-
ble improvements—often with notable deteriorations—in employment
and income distribution. The revolution happened just when Cuba was
on the brink of playing an important part in the incipient regional restructuring. On the eve of the 1960s, given its special relationship with
the United States, relative development, and regional prominence, Cuba
would surely have played a central role in these transformations.

By virtue of the revolution, however, Cuba embarked upon a markedly distinct development path from the rest of Latin America. In principle, socialism allowed the state to pursue more rational economic policies responsive to Cuban interests. Satisfying the basic needs of the *clases
populares* was now the principal objective. Revolutionary leaders and
foreign observers alike predicted spectacular growth rates and rapid improvements in living standards.[2] The 1950s had also witnessed growing
optimism about the prospects for Cuban development. Felipe Pazos had
often drawn parallels between Cuba and Norway and Denmark.[3] "Cuba
is not a typical underdeveloped country, but rather . . . an imperfectly
developed nation," he had affirmed.[4] In 1954, the U.S. economist Byron
White had written a book comparing Cuba and Denmark.[5] In 1956, the
U.S. Department of Commerce had noted that Cuba was not an underdeveloped country in the usual sense.[6] The economic thesis of the July
26th Movement had likewise professed the belief that Cuba was potentially wealthy and needed only appropriate policies to reap prosperity.[7]
These assessments, however, had never contemplated a break with capitalism.

Nonetheless, the revolution bore the burden of extraordinary expectations. The elimination of capitalism was then thought to enhance economic growth as well as social justice. Revolutionary leaders strongly
subscribed to the notion that the new directions were sufficient to overcome the gap between Cuban potential and reality. As the 1960s began,
when worldwide assessments for development were inordinately optimistic, Cuba appeared to be in a particularly favorable position. Socialism uncovered new vistas and seemed to foster the right directions. The
outcome of dependent socialist development, however, cast a shadow of
realism over the unbounded perspectives entertained by the *rebeldes,* the
Cuban people, and many foreign sympathizers more than three decades
ago.

The past—sugar monoculture and U.S. dependence—burdened the
new state more heavily than anticipated. The United States did not
accept Cuban self-determination and after 1962 imposed a steep over-
head by means of an embargo. Socialism—based on inclusive development and close ties to the Soviet Union—constrained Cuban development in ways unexpected during the heyday of revolution.[8] Moreover,
after 1989, the collapse of Eastern Europe and the disintegration of the

Soviet Union undermined the viability of Cuban socialism. Nonetheless, Cuban development trends and the subsequent record of dependent capitalism elsewhere in Latin America suggested capitalism might have likewise—albeit costs and benefits would have accrued to different social sectors—disappointed the glowing expectations of national prosperity that had abounded during the 1950s.

During 1959–1960, the Cuban economy performed well. The end of civil strife and the maturity of investments made during the 1950s contributed to economic recovery. Growth rates were probably about 10 percent a year.[9] In 1959–1961, sugar output averaged about 6.2 million tons annually, an improvement over the 1950–1958 average of 5.4 million.[10] The 1954–1958 trade trends were reversed when slightly lower deficits were recorded in 1959 and when a surplus of more than 28 million pesos was achieved in 1960.[11] Thus, the revolutionary government had the resources to grant immediate benefits to the *clases populares* and muster their support against the domestic opposition and the United States. During the first eighteen months after January 1, 1959, no less than 15 percent of national income shifted from property owners to wage earners.[12] Socialism, moreover, promised to accelerate these gains. An aura of rationality then enveloped economic planning and reinforced the conviction that development would follow almost automatically. The government forged the industrialization strategy of 1961–1963 in an ambience of optimism.

The objectives of the 1961–1963 strategy were reasonable and their rationale not unfamiliar to reformers in the old Cuba. Sugar production was to expand to about 7 million tons annually. Although the volume of sugar exports would be maintained, the sugar share of total exports was expected to decline from 80 percent to 60 percent. Agricultural diversification would allow for greater food self-sufficiency. Import-substitution industrialization in metallurgy, transportation equipment, chemical products, and machinery would provide the inputs to modernize agriculture. The strategy also targeted the industrial development of cane by-products. As during the 1950s, economic prospects depended on finding and exploiting large petroleum reserves. Credits from the Soviet Union and other socialist countries, funds from previously remitted profits, and savings from luxury imports would finance investment. Socialist aid was considered a transitory necessity to spur development and attain a favorable trade balance.

The strategy of rapid industrialization failed. It did not fully consider the costs and levels of imports needed to implement import-substitution. In 1962–1963, partially because the amount of land used to grow cane was reduced, sugar output declined precipitously to 4.8 and 3.8 million tons.[13] Moreover, agricultural diversification did not satisfy domestic demands for food or generate sufficient exports to cover the sugar shortfall. Thus, the trade deficit seriously deteriorated: a deficit of 12.3 million

pesos in 1961 increased to 237 million in 1962 and 322.2 million in
1963.[14] Contrary to expectations, central planning was often improvised
and chaotic and did not provide quick solutions to economic problems.
Emigration of professionals and skilled workers aggravated the relative
scarcity of technical and administrative personnel. The U.S. embargo
and the weather also hindered the initial strategy. In 1964, the revolutionary government abandoned industrialization and adopted a "turnpike" strategy centered on sugar and agriculture.[15]

The failure of rapid import-substitution industrialization underscored the difficulties in overcoming monoculture and external dependence. Self-sustained, balanced development was now a long-term objective. Industrialization and diversification depended on foreign
exchange earnings that only the sugar sector could realistically command. The sugar agro-industrial complex was to be the fundamental
engine of growth. Establishing forward and backward linkages between
the sugar sector and other economic activities was pivotal to the new
strategy. Increasing production in other agricultural sectors such as cattle, citrus fruits, tobacco, and coffee would complement the turnpike
strategy. Producing a sugar harvest of 10 million tons in 1970 became the
medium-term objective.[16] The Soviet Union committed credits and
guaranteed markets at modestly preferential prices to underwrite the
sugar-centered drive.

Although the 1970 harvest did not produce 10 million tons, the
post-1964 strategy was not discarded. The state continued to emphasize
sugar as the principal source of foreign exchange and its modernization
as the center of backward and forward linkages. In 1972, Cuba became a
member of the then-socialist trading bloc Council for Mutual Economic
Assistance (CMEA) and obtained more favorable arrangements for its
sugar exports, preferential prices for nickel exports, loans at low interest
rates, a fifteen-year postponement of debt repayment, and below-world-
market prices for petroleum imports.[17] CMEA membership, however,
also reinforced the role of sugar in the Cuban economy.

Athough the collapse of socialism in Eastern Europe and the disintegration of the Soviet Union radically altered the international conditions sustaining the Cuban economy, Cuban economic relations with
the then-socialist countries were changing and offered diminished perspectives even before 1989. During the early 1980s, the Soviet Union was
already reducing its special relationship with Cuba. CMEA countries
were emphasizing efficiency and cost-accounting in determining the
terms of trade.[18] Total imports from the Soviet Union and Eastern Europe were fluctuating within declining rates of increases.[19]

Dependent socialist development resulted in a mixed record. During
the 1960s, the economy registered a slight decline per capita. The 1970s
were the only period of sustained economic growth, about 5 percent per
capita. Although per capita growth was estimated at no less than 2.5

percent during the early 1980s, a downturn of more than − 4 percent per capita in 1987 augured the calamitous contraction that was to follow. In 1988–1990, growth was negligible; in 1991–1992, economic decline was said to have been no less than 35 percent.[20] The state emphasis on equity, however, partially protected the *clases populares* from the consequences of inconsistent growth. Since 1959, the share of national income going to the bottom 40 percent of the population had increased from 6.5 percent to about 26 percent.[21] Nonetheless, the centrality of sugar continued to hamper economic growth and limit living standards.

During the 1980s, sugar averaged 76.6 percent value share of total exports, only moderately lower than in the 1950s.[22] Per capita sugar output, however, revealed an enduring dilemma: Cuba was producing about .7 ton of sugar per person—only 17 percent higher than the trough of the 1930s, and 10 percent and 23 percent lower than in the 1940s and 1950s, respectively.[23] Modest export diversification did not compensate for a declining per capita sugar output, for which there was still a stagnant and even contracting international market.[24] The impressive modernization of production poignantly underscored the dilemma. Among sugar cane exporters, Cuba had the most advanced infrastructure. The sugar industry had, moreover, established linkages to other economic sectors: its modernization had created a small capital-goods sector and the development of sugar by-products had also progressed modestly.[25] Sugar exports, however, were never going to sustain economic growth. Moreover, the collapse of the preferential terms of trade between Cuba and the former Soviet Union further underscored the definitively doomed prospects of sugar as an engine for development.

A sugar-centered export sector, however, obscured the moderate progress in diversifying the domestic economy after 1959. Although during the 1950s total trade had averaged 55 percent of Gross National Product (GNP) and exports 28.9 percent, the respective proportions of Gross Social Product (GSP) in the 1980s were 48.5 percent and 21.1 percent.[26] Sugar now accounted for 16.8 percent of industry; 29.6 percent of agriculture; 9.5 percent of GSP; and 9.3 percent of the labor force.[27] The comparable figures in the 1950s had been 34.5 percent of industry; 55.8 percent of agriculture; 28 percent of GNP; and 23 percent of the labor force.[28] The changed structure of imports was another indication of discreet transformation. During the 1980s, consumer goods averaged 11.4 percent of imports; intermediate goods 65.6 percent; and capital goods 23 percent.[29] During the 1950s, consumer goods had been 41.0 percent of imports; intermediate goods, 34.8 percent; and capital goods, 24.2 percent.[30] As a proportion of total imports, moreover, food imports had declined almost 50 percent (see Table 4.1).

Although three decades after the revolution Cuba appeared to be more self-sufficient in food, the lower share of food imports was misleading. Because of the post-1959 modernization, agriculture was more

Table 4.1. Monoculture and Dependence, Cuba, 1950s and 1980s
(in percentages)

	1950s	1980s
Sugar exports/Total exports	83.0	76.6
United States/USSR		
Total trade	68.2	66.2
Sugar exports (values)	54.8	75.8
Total Trade		
GNP	55.0	
GSP		48.5
Exports		
GNP	28.9	
GSP		21.1
Imports		
Consumer goods	41.0	11.4
Intermediate goods	34.8	65.6
Capital goods	24.2	23.0
Sugar Shares		
Industry	34.5	16.8
Agriculture	55.8	29.6
GNP/GSP	28.0	9.5
Labor force	23.0	9.3
Sugar tonnage per capita	.86	.70
Decade trade surplus/deficit per capita in pesos	+61.2	−1300.0

Sources: Oscar Zanetti, "El comercio exterior de la república neocolonial," *Anuario de estudios cubanos: la república neocolonial* (Havana: Editorial de Ciencias Sociales, 1975), pp. 78, 115; Comité Estatal de Estadísticas, *Anuario estadístico de Cuba, 1986,* pp. 192, 235, 297, 406–408, 422–425, 461–463, and *Anuario estadístico, 1988,* pp. 57, 99, 235, 243, 300, 410–412, 426–427, 429, 467; Hugh Thomas, *Cuba: The Pursuit of Freedom* (New York: Harper & Row, 1971), pp. 1563–1564; William M. LeoGrande, "Cuban Dependency: A Comparison of Pre-Revolutionary and Post-Revolutionary International Economic Relations," *Cuban Studies/Estudios Cubanos* 9 (July 1979): 6, 14; Carmelo Mesa-Lago, *The Labor Force, Employment, Unemployment and Underemployment in Cuba: 1899–1970* (Beverly Hills: Sage, 1972), pp. 23, 29; Claes Brundenius, *Revolutionary Cuba: The Challenge of Economic Growth with Equity* (Boulder: Westview Press, 1984), p. 147; Banco Nacional de Cuba, *Memoria, 1958–1959* (Havana: Editorial Lex, 1959), pp. 189–191.

dependent on imports, especially fuel. Moreover, the food plan of 1990 aiming for self-sufficiency only highlighted the past failure of the government in this basic economic task. The decline in the import of consumer goods responded to the priority of satisfying basic needs and directing national resources toward productive investment. The large proportion of intermediate-goods imports underscored the trade-dependent effort to industrialize and modernize agriculture in a country with limited raw materials and energy resources. The similar import share of capital goods highlighted strong dependence on external sources of capital in both periods.

Sugar, then, continued to be the bulwark of Cuban dependence. During the 1980s, the Soviet Union accounted for 66.2 percent of total trade; during the 1950s, the United States, 68.2 percent. Because the Soviet Union bought Cuban sugar at high preferential prices, its value share of sugar exports was 75.8 percent; its volume share was 56.3 percent. During the 1950s, the United States had bought 54.8 percent of Cuban sugar exports.[31] That Cuba during the 1950s and 1980s traded largely with a single partner that received a majority of sugar exports underscored the openness of the Cuban economy. Moreover, the enduring centrality of sugar aggravated the balance of trade. During the 1950s, terms of trade had been rapidly declining relative to the late 1940s, when Cuba had accumulated nearly 1.4 billion pesos in trade surpluses (280 pesos per capita). Between 1950 and 1958, trade surpluses plummeted to about 367 million pesos (61 pesos per capita).[32] After 1959, Cuba obtained modest trade surpluses only in 1960 and 1974. Between 1959 and 1969, total deficits were about 2.7 billion pesos (432 pesos per capita). The situation in the 1970s improved slightly: in current pesos, deficits totaled a little over 3 billion pesos (326 pesos per capita). However, deficits more than quadrupled, to about 13 billion pesos (1300 pesos per capita), between 1980 and 1988.[33]

During the 1980s, trade deficits with the socialist countries increased but declined in relation to the total deficit. Between 1983 and 1985, hard-currency deficits grew faster than those with the CMEA. In 1986, Cuban deficits with the Soviet Union and Eastern Europe again rose relative to the total deficit.[34] In the mid-1980s, hard-currency earnings contracted more than 50 percent, and therefore trade with market economies declined.[35] Total debt completed the sobering panorama of continued dependence. As of 1990, Cuba owed the Soviet Union 15.5 billion rubles, and hard-currency creditors nearly $7 billion.[36] Hard-currency debt represented about $700 per capita, and payments absorbed nearly 60 percent of the value of convertible currency exports.[37] In May 1991, Cuba had hard-currency reserves of less than $84 million.[38]

Often disappointing performance and the barely changed dependence on sugar exports still defined the Cuban economy. Quantitative values, however, masked the qualitatively different consequences that dependence on the United States before 1959 and on the Soviet Union subsequently brought to Cuban society. The domestic preeminence of sugar had been modestly reduced. Moreover, in spite of lower per capita sugar outputs and diminished prospects in the international market, socialist Cuba attained significant advances in the satisfaction of basic needs. Nonetheless, as the National Bank had so starkly forecasted in 1956, continued dependence on sugar exports was constraining living standards. Relative equality was possible because of the radical transformation of class relations: the Cuban state was committed to the *clases populares*, and dependent socialist development had facilitated meeting their basic needs.

The uncertainties of the early 1990s, however, raised new concerns. Trade with the former Soviet Union declined more than 50 percent and was no longer conducted on preferential terms.[39] In October 1992, the signing of a new trade agreement between Russia and Cuba seemed to signal a modest recovery though not under previous terms.[40] In 1990, cutbacks in oil deliveries forced the Cuban government to proclaim a "special period in peacetime," adopting wide-ranging austerity measures and predicting hard times for the foreseeable future. Modestly increasing trade with China, North Korea, Iran, and Latin America offered some respite. Tourism was rapidly expanding: it generated more than $400 million in gross revenues in 1992. The Cuban government, moreover, was having measured success in attracting foreign capital in joint ventures to build hotels. Foreign investments in other economic sectors, however, were not yet significant.[41] Although the government hoped to export biotechnological products to Third World countries, their outlook as an important source of hard currency was quite modest.[42]

None of these prospects amounted to a substitute for the generally favorable relationship Cuba had had with the Soviet Union and Eastern Europe. The United States, moreover, remained as staunch an opponent of the Cuban government as ever. In October 1992, George Bush signed the Cuban Democracy Act, which prohibited U.S. subsidiaries abroad from trading with Cuba. Thus, Cuba had to restructure its international economic relations almost as substantially as it had done during the early 1960s but under even more inauspicious conditions. Then, the promises for development seemed bright and the socialist countries an alternative source for trade, credits, and aid. Now, the likelihood of sustained economic growth was quite dim, and there were no alternatives to participating in the capitalist world economy. The unanticipated conditions of the early 1990s likewise threatened Cuban accomplishments in the satisfaction of basic needs. Prospects for maintaining the austere living standards of the *clases populares* were not promising. Indeed, without the Soviet Union, Cuban socialism and the premise of inclusive development looked to be economically untenable.

Standards of Living after the Revolution

The problem of employment was an immediate priority for the revolutionary government. During the 1960s, seasonal unemployment was eliminated and almost all working-age Cubans found stable employment. Between 1962 and 1969, an unofficial estimate of unemployment was 4.7 percent, well below the level of the 1950s.[43] In 1970, annual unemployment was 1.8 percent.[44] Unofficial unemployment estimates for the 1970s were between 2.4 percent and 3.8 percent.[45] In 1981, the census recorded a 3.4 percent unemployment rate. On average, unemployment in the Oriente provinces was 4 percent, and lower than the national average in the other five provincial groups.[46] By the 1980s, the significant changes in the EAP were in agriculture (down, from 41 to 20 percent) and services (up, from 20 and 30 percent).[47]

Low levels of unemployment, however, were not entirely salutary because there were no official data on underemployment. Both censuses after the revolution classified as "employed" all persons working full-time, part-time for at least fifteen hours a week, and without pay for a relative.[48] During the 1950s, the latter two types of work had constituted underemployment. Full employment thus masked considerable under-employment, as evidenced by chronically low labor productivity, and underscored a delicate predicament for a state committed to economic growth and social justice. Moreover, the economic downturn of the late 1980s and early 1990s was forcing numerous layoffs and plant shutdowns. Unemployment was growing, and the Cuban government was therefore faced with breaching a central tenet of its commitment to the *clases populares*.

During the 1950s, a majority of Cubans had lived in urban areas. In 1970, the urban population (60.3 percent) barely surpassed the levels of 1953. Stagnant urbanization, however, concealed an important transformation: between 1953 and 1970, the number of urban centers with over 20,000 persons grew and their inhabitants notably increased as a proportion of the urban population. In 1981, the census revealed a somewhat different trend: overall urbanization (69 percent) increased more significantly, but urban concentration did not. Slightly lower proportions of the urban population then lived in centers of more than 20,000 persons. During the 1960s, urbanization stagnated while urban concentration expanded. During the 1970s, Cuba registered significant urban growth from the bottom up. Both trends pointed to a markedly more urban nation. Between 1953 and 1981, moreover, the city of Havana's slightly declining share of the total population (from 20.9 percent to 19.8 percent) underscored a pattern of less skewed urbanization.[49]

After 1959, educational levels improved markedly. By the 1980s, literacy was almost universal and virtually uniform throughout the island. School enrollment for 6- to 14-year-olds also registered remarkable uniformity around the national average of more than 95 percent; those for 6- to 24-year-olds were somewhat higher in urban areas.[50] By 1981, only 3 percent of the 6- to 49-year-old cohort reported no schooling, and more than two-thirds were at or above the sixth-grade level.[51] About 6 percent of the population 6 years and older was either in high school or had graduated from high school; students and graduates from universities were 3.3 percent. Generally, the population in urban areas, especially the Havana provinces, had higher educational attainments.[52] Thus, regional differences were significantly reduced (see Table 4.2).

Socialism also improved national health care. Before the revolution Cuban aggregate health indices had been among the best in Latin America, but these ensconced profound urban-rural differences, especially between Havana and the rest of the country. After 1959, aggregate indices more accurately reflected the state of public health and the distribution of health services. During the 1980s, Cubans died most often of cardiovascular diseases, malignant tumors, and cerebrovascular diseases.[53] The first two, however, had also accounted for two of the top three causes of death before 1959. Life expectancy increased to about 75 years.[54] Crude death rates remained basically unchanged over the three decades since 1959.[55] During the 1960s, infant mortality actually increased—whether in fact or because of better reporting—and subsequently declined to 10.7 percent in 1990.[56] Although the Havana provinces continued their preferential status, provincial distribution of doctors, hospital beds, and health care units registered substantial improvements. About 46 percent of the more than 31,000 doctors (1:333 persons), and nearly 40 percent of all hospital beds were in the

Table 4.2. Selected Indicators, Cuba, 1980s (in percentages)

	Urban	Rural	Havana[a]	National
School enrollment, 6- to 24-year-olds[b]	75.4	67.2	73.7	72.6
6- to 49-year-old cohort at or above 6th grade	n.a.	n.a.	75.5	66.8
Some university education[b]	4.4	2.0	6.0	3.0
Health care[c]				
Doctors	n.a.	n.a.	46.4	53.6
Hospital beds	n.a.	n.a.	38.9	61.1
Health facilities	n.a.	n.a.	24.2	74.3
Housing units[c]	n.a.	n.a.	25.7	74.3
Total wages[c]	n.a.	n.a.	34.4	65.6
1981–1988 investments	n.a.	n.a.	32.9	67.1
State civilian construction	n.a.	n.a.	37.7	62.4

Sources: República de Cuba, *Censo de población y viviendas de 1981*, vols. 2, 3, 16, pt. 2, pp. 160, 169, 163, 150–153, 156; Comité Estatal de Estadísticas, *Anuario estadístico de Cuba, 1982*, p. 191, and *Anuario estadístico de Cuba, 1988*, pp. 191, 197, 223, 279, 282, 563–564, 567, 571.

[a] City of Havana and Havana provinces.

[b] 1981.

[c] 1988. National = all other provinces.

Havana provinces—a reflection of the vertical organization of the health care system. Primary medical care was more evenly and widely available in all provinces and regions.[57] Health care, moreover, constituted a point of contact between the citizenry and the government that the leadership accorded the utmost importance.[58]

Housing presented a decidedly mixed performance.[59] By the 1980s, the housing deficit ranged between 1.2 to 1.5 million units. Housing quality was also inadequate. In 1981, nearly two-thirds of all housing was considered solidly constructed: about three-quarters of urban housing and over a fifth of rural.[60] That only about 55 percent of post-1959 housing was solidly constructed revealed the gravity of the situation.[61] Housing construction, nonetheless, gave evidence of government efforts to mitigate urban-rural differences: more than two-thirds of all rural housing dated after 1959, and about two-fifths of urban quarters did.[62] Between 1982 and 1988, the distribution of completed housing units generally corresponded to provincial shares of the total population.[63] Moreover, significant progress was made in the proportions of all living quarters with inside plumbing (53 percent), an inside toilet (45 percent), electricity (83 percent), and a shower (52 percent). Urban areas retained their advantages over the countryside, albeit less glaringly than before 1959.[64] The distribution of some household goods also increased. By 1981, four in five households had a radio. Nearly three in four households in urban areas and one in four in the countryside had a television set. About 65 percent of urban housing and 18 percent of rural housing had a refrigerator. More than half of urban households and more than two-fifths of rural ones had a sewing machine.[65]

During the 1950s, development trends had aggravated provincial inequalities, especially those between Havana and the rest of the island. The revolution arrested these trends, but Havana continued to be pre-eminent.[66] During the 1980s, the distribution of total wages was much more equitable. The Havana provinces accounted for about 35 percent. Only the five Oriente provinces registered a wage share (about 27 percent) significantly below their population share (about 36 percent). Between 1980 and 1988, the average annual salary was somewhat higher than the national average in the Havana (3.4 percent) and Las Villas provincial groups (1.5 percent); about the same in Matanzas and Camagüey/Ciego de Avila; and approximately 5 percent below in Pinar del Río and the Oriente provinces.[67]

If somewhat more equitably, average 1981–1988 investments showed similar distributions. The Havana and Oriente provinces received 32.9 percent and 27.7 percent of state investments. Matanzas (7.3 percent) and the Las Villas provinces (16.3 percent) followed Havana in greater proportions of total investment relative to their share of total population. Camagüey/Ciego de Avila (10.4 percent) received approximately their fair share, and Pinar del Río (4.5 percent) was slightly underrepresented in total investments.[68] The value distribution of state civilian construction again underscored progress toward bridging regional differences. The Havana provinces represented 37.7 percent of all construction, and the Oriente group accounted for 26.7 percent. Matanzas (6.6 percent) and Las Villas (16.0 percent) received higher proportions relative to their share of total population, Pinar del Río (5.2 percent) and Camagüey/Ciego Avila (8.3 percent) somewhat lower. The widest gap was once again between the Havana and Oriente provincial groups.[69] Nonetheless, overall trends highlighted the relative success of state policies in promoting less unequal national development.

At the beginning of the 1990s, Cuba had one of the most skilled labor forces in Latin America. At least 35 percent had graduated from high school, vocational training, teacher training, or higher education. More persons in the labor force in Havana and Oriente were similarly skilled: 41.6 percent and 38.8 percent, respectively.[70] Thus, Cuba had a considerable resource in human capital. More salient then were questions regarding the underutilization of these skills. Indeed, the Cuban government had not met the challenge of translating these impressive human capital investments into sustained increases in labor productivity and economic growth.

Socialist Visions and Inclusive Development

Although socialism was an alternative that Cuban history and the social revolution of 1959 made possible, the Cuban leadership emphasized the role of subjective factors in making that possibility a reality. The *fidelista* historical experience constituted a pivotal point of reference for the focus on forging a new *conciencia*, and thus emphasis on human will—often over structural realities—was one of the enduring trademarks of development policies. In 1960, Fidel Castro had emphasized revolutionary *conciencia* as more powerful than gold for the overwhelming majority of Cubans who had affirmed their *patria* and upheld the interests of the *clases populares*. Massive popular support was an immeasurable resource. Past memories, present reality, and future promises had ignited the *clases populares* in 1959–1960 and mobilized them during Playa Girón and the Missile Crisis. For three decades, state development policies endeavored to balance economic and moral goals. Competing visions of socialism buttressed different approaches to planning and incentives. Even so, the intransigence of radical nationalism on national sovereignty and social justice ultimately proved to be the compass of Cuban socialism.

The early 1960s were years of extraordinary euphoria and high expectations. Success against Batista and the old Cuba endowed the *rebeldes* with a fanciful sense of their own possibilities for composing the new Cuba. Failure to attain rapid growth rates and economic diversification, however, dispelled the notion that socialism would be a quick panacea. The popular dictum—*Cuba país rico, pueblo pobre* (Cuba a wealthy country whose people are poor)—that the revolution appropriated in its early prognostications receded with the realization that Cuba could overcome underdevelopment only in austerity. Indeed, the very magnitude of the task challenged the leadership to battle. While rapid industrialization had failed, socialism would succeed nonetheless. The Great Debate of 1962–1965 addressed crucial issues on the structure of planning and incentives. By then, Cuban leaders knew economic prospects were limited and would be so for the foreseeable future.

Industry Minister Ernesto Guevara succinctly raised the question that set the tenor of the debate: "How, in a country colonized by imperialism, its basic industries underdeveloped, a mono-producer dependent on a single market, can the transition to socialism be made?"[71] The central issue turned on the role of the law of value in the Cuban economy. The revolution had broadened the potential for consumption of all sectors of society and, consequently, demand outstripped economic capacity. Because inclusive development precluded widening inequalities and marginalizing the *clases populares*, how should socialism regulate production, accumulation, and distribution? Under capitalism, profit maximization guided investments. In underdeveloped countries, the ensuing economic and social unevenness of dependent capitalism was particularly marked. Classical Marxism identified the essence of communism to be the elimination of the law of value and market relations. Socialism aimed to move in the direction of curtailing wage labor so that people could progressively satisfy their needs on the basis of cooperation. How Cuba could best advance socialist visions was the crux of the Great Debate. Over a period of three years, two positions emerged.[72]

One response to Guevara's question about the transition to socialism in Cuba adhered to an orthodox interpretation of the relationship between material development and social consciousness. The mature Marx unequivocally expressed a determinist interaction:

> In the social production of their existence, men inevitably enter into definite relations, which are independent of their will, namely relations of production appropriate to a given stage in the development of their material forces of production. The totality of these relations constitutes the economic structure of society, the real foundation, on which arises a legal and political superstructure and to which correspond definite forms of social consciousness.[73]

Hence, Cuban underdevelopment required a correspondence between economic organization and the level of available technology and skills. Recognition of the law of value was necessary to develop the economy. Less centralized planning and moderate use of market relations and material incentives could best regulate value, productivity, and efficiency. Albeit limited by state ownership of the means of production, profitability was the most viable criterion to guide production, accumulation, and distribution. Alternate criteria would have to be defined administratively, which carried the dangers of a bureaucratic morass, economic inefficiency, and the breakdown of planning. Cuba lacked the skills to enforce highly sophisticated central planning. Moreover, the *conciencia* of daily life differed from that of Playa Girón and the Missile Crisis, and recompense for work had to recognize these differences. Costly consequences could well follow the excessive curtailment of the law of value in a small country like Cuba with limited natural resources and extraordinary dependence on foreign trade. *Conciencia* could simply not be divorced from the economy and the level of development. INRA President Carlos Rafael Rodríguez and Foreign Trade Minister Alberto Mora were the foremost proponents of this position.

Ernesto Guevara and others contended that Cuba could not allow the law of value to determine investments without reneging on the possibility of overcoming underdevelopment. Industry did not enjoy the comparative advantage of agriculture and, therefore, was not as "profitable." Self-finance planning would tend to reinforce uneven development and specialization. The budgetary system of centralized planning allowed the state to plan for the economy as a whole, correct past inequalities, and promote more balanced development. The fact that Cuba was a small country with limited wealth and an open economy compelled the state to harness its most abundant resource: the will, energy, and passion of the Cuban people. Self-finance planning advocated material incentives on the grounds of efficiency and rationality. Yet material incentives privatized *conciencia*, and inefficiency was not restricted to economic resources. Moral incentives would develop *conciencia* as an economic lever and further the creation of new human beings. Guevara forcefully asserted:

> Pursuing the chimera of achieving socialism with the aid of the blunted weapons left to us by capitalism (the commodity as the economic cell, profitability, and individual material incentives as levers, etc.), it is possible to come to a blind alley. And the arrival there comes about after covering a long distance where there are many crossroads and where it is difficult to realize just where the wrong turn was taken. Meanwhile, the adapted economic base has undermined the development of consciousness. To build communism, a new man must be created simultaneously with the material base.[74]

These two visions of socialism and their concomitant models of economic organization guided state development policies in varying ways and at different times. During the early 1960s, budgetary planning operated in industry and self-financing in agriculture. During the late 1960s, a version of budgetary financing prevailed as Cuba launched a radical strategy of centralization, moral incentives, and mass mobilizations. During the 1970s and early 1980s, a version of self-financing was operant in the economic management and planning system (SDPE). After 1986, a partial retrenchment from the SDPE and a renewal of the moral dimension of socialism took place. At each turn, Cuba confronted an assortment of domestic and international factors that influenced the relationship between state and market. The availability of labor, the economic feasibility of material incentives, the tension between growth and equity, the weight of ideological perspectives, the flow of hard currency, the disposition of the Soviet Union and the other socialist countries to aid Cuba, and the continuing burden of the U.S. embargo were among these factors.[75]

Post-1959 Cuban development was based on altered class relations and restructured international constellations. The consequences, costs, and benefits of inclusive development were markedly different from those that would have followed had the transformation of dependent capitalism continued along the directions of the 1950s. The social revolution extended the modern profile of the old Cuba and affirmed independence from the United States. Nonetheless, after three decades, the model of dependent socialist development that in many ways had served Cuba well had run its course. International and domestic conditions undermined the viability of Cuban socialism. During the early 1990s, the gap between official discourse, economic policies, and the expectations of the citizenry was rapidly widening.

Cuban leaders never viewed democracy independently from radical transformation and national self-determination. For three decades, they addressed questions of democracy within the vanguard party model of politics. Each decade provided a different set of answers. Economic and political organization followed a complementary, if often tense, development. The three-decade trajectory of the Cuban Communist Party, the Central Organization of Cuban Trade Unions, and the Federation of Cuban Women highlighted the interaction between politics and economics.

CHAPTER 5

Politics and Society, 1961–1970

The institutionalization of the revolution has yet to be achieved. We are seeking something new that will allow a perfect identification between the government and the community as a whole, adapted to the special conditions of socialist construction and avoiding to the utmost the commonplaces of bourgeois democracy transplanted to the society in formation. . . . We have been greatly restrained by the fear that any formal aspect might make us lose sight of the ultimate and most important revolutionary aspiration: to liberate man from alienation.

Ernesto Guevara
March 1965

We shall seek our own revolutionary institutions, our own new institutions, stemming from our conditions, from our idiosyncrasy, from our customs, from our character, from our spirit, from our thought, from our creative imagination. We shall not imitate.

Fidel Castro
September 1965

Charismatic authority and popular mobilizations crystallized the politics of the new Cuba: Fidel Castro was fulcrum; *el pueblo cubano*, sustenance. Having rejected representative democracy, Cuban leaders confronted the challenges of governance. Maintaining elite unity and mobilizing popular support were their core concerns. Thus, the revolution brought together the July 26th Movement, the Revolutionary Student Directorate, and the Popular Socialist Party in a vanguard party. In 1965, after considerable conflict between new and old communists, the Cuban Communist Party was formed. The Central Organization of Cuban Trade Unions and the Federation of Cuban Women were two of the mass organizations involving ordinary Cubans in the tasks of socialism. The early 1960s witnessed an incipient institutional order.

By middecade, Cuban leaders reconsidered the early process of institutionalization. Political and economic factors—both domestic and international—convinced them that the models borrowed from the Soviet Union and Eastern Europe were undermining the revolution. Socialism was not summoning the popular enthusiasm that the social revolution had. Austerity—not standards of living comparable to Western Europe—marked daily routines. The revolutionary leadership sought to establish greater political and economic independence from the Soviet Union. Between 1966 and 1970, Cuba attempted to pursue the parallel construction of communism and socialism: a radical experiment to develop *conciencia* and the economy simultaneously. Cuban leaders hoped to generate sufficient resources to allow them a more balanced relationship with the Soviet Union and to institutionalize the revolution using their own models.

The Incipient Institutional Order, 1961–1965

Politics in the new Cuba required the mobilization of *el pueblo cubano* in defending the nation and developing the economy. By 1961, popular militias numbered more than 300,000; the Committees for the Defense of the Revolution, nearly 800,000.[1] The Federation of Cuban Women mobilized housewives for the chores of the revolution. The trade unions sought a new profile—more subordinate, less "economistic"—under socialism. Born in urban Cuba, volunteer work spread to the countryside. Thousands of city dwellers wielded *machetes* (cane knives), downed the tall stalks of cane, and had their first encounters with the reality of underdevelopment. The literacy campaign enlisted nearly 300,000 people.[2] The struggle to survive infused these mobilizations with a military mission. Institutionalizing the revolution, however, supposed a civilian understanding of participation or, at least, involvement and consultation. Tensions between military exigencies and civilian imperatives punctuated the new politics.

Popular mobilizations supported charismatic authority. Ernesto Guevara often praised the appearance of *las masas* in the revolution: the masses were a conscious people forged in the heat of the agrarian reform, Playa Girón, the Literacy Campaign, the Missile Crisis, the daily construction of socialism. *Las masas* followed Fidel without vacillations but with convictions. Fidel was a *maestro:* the people trusted him because he expressed their wishes and needs. Communication between Fidel and *el pueblo* was like "a dialogue of two tuning forks whose vibrations summon forth new vibrations . . . of growing intensity . . . crowned . . . by struggle and victory."[3] Together, they had defeated the United States and the *clases económicas.* The nation was up in arms under the leadership of the *Comandante en Jefe* (commander in chief).

The revolution was also socialist, and socialism supposed a central role for the working class. Workers as members of *las masas* enthusiastically supported the revolution; as members of the working class, they expressed more erratic allegiance. Defending *la patria* was a matter of national honor. Working under state management was not as heroic and generated considerable more conflict. Daily life and work called for the mediation of institutions; governing required a routine of decision making. And more normal times did not easily incorporate the exhilaration of a social revolution. Cuban leaders viewed the routinization of politics with suspicion: institutions buffered communication with the people. The early 1960s confirmed their misgivings and subsequently led them to seek ways of retaining the effervescence of revolution.

The Formation of a Vanguard Party

During 1959–1960, the *rebeldes* proved to be quite adept in restraining organizations against the revolution. Their experience in forging revolutionary organizations was, however, more limited. The struggle against Batista had emphasized military and clandestine skills. The first two years in power had also underscored mobilization against enemies rather than discourse in the give-and-take of politics. Uniting the three organizations that continued to function at the end of 1960—the July 26th Movement, the Revolutionary Student Directorate, and the PSP— became imperative as the revolution closed ranks against the domestic opposition and the United States.

The model of a vanguard party was compelling for two reasons. First, the experience of the socialist countries gave testimony to the efficacy of vanguard parties in consolidating and retaining power. When the revolution became socialist, the notion of a vanguard organization as the cornerstone of the new politics was reinforced. Planned economies and one-party polities seemed to be natural complements. Moreover, severed ties with the United States necessitated finding new markets and protection against aggression, and the Soviet Union provided them. Without a vanguard organization, Cuba would not be recognized as a bona fide member of the socialist community.[4] Second, the Cuban revolutionary tradition offered precedent to underwrite the vanguard concept. José Martí in the 1890s and Antonio Guiteras in the 1930s had advocated a united vanguard to advance the popular cause.[5] More recently, Fidel had demonstrated the value of unity in the overthrow of Batista. Just as important, Cuban history offered numerous examples of disunity undermining effective political action at crucial moments. The independence movement against Spain and the short-lived Grau-Guiteras government of 1933–1934 had been two crucial instances. Thus, the revolution appropriated the then-socialist model of politics.

During the summer of 1961, the July 26th Movement, the DRE, and the PSP dissolved and formed the Integrated Revolutionary Organizations. Nonetheless, only the PSP had the cadres needed to organize a party. The July 26th Movement and the DRE were of recent founding, lacked infrastructures, and had depended on the leadership of Fidel Castro and José Antonio Echevarría. Moreover, the course of events during 1959–1961 and the establishment of socialism had raised the importance of the PSP for the revolution. After all, the PSP knew about socialism, vanguard parties, and the Soviet Union. The July 26th Movement and the Revolutionary Student Directorate did not.

When Aníbal Escalante became organization secretary, the PSP assumed control of the ORI. Escalante used the PSP infrastructure to organize the new party and therefore tended to favor the communists over the July 26th Movement and the DRE. Because recruitment was based on membership in the July 26th Movement, the DRE, and the PSP, mass participation in the selection process was, moreover, precluded. Although within a few months the ORI had about 15,000 members, party formation was exacerbating deep-seated tensions among the three organizations and was divorced from *el pueblo cubano*.[6] Thus, the two objectives the leadership hoped a vanguard party would meet—elite unity and popular involvement—were being sidetracked.

In March 1962, Fidel Castro accused the ORI leadership of sectarianism, abuse of authority, and disdain for the people. Charging that the new party was causing popular disenchantment, Fidel observed: "We were witnessing a veritable loss of faith in the revolutionary leadership. . . . [T]he masses had more sense of what was going on than the revolutionary cadre. . . . [W]e were living in an ivory tower . . . actually we had lost contact with the masses."[7]

As styles and attitudes reminiscent of the past proliferated, the revolutionary government was becoming a "parody." The motto ¡*ORI es la candela!* (ORI is the one!) sung to the tune of a conga unabashedly evoked the demagoguery of the old *políticos*. Nonetheless, Fidel insisted a party organized along "true Marxist-Leninist principles" was needed.[8] One of these principles was a new recruitment method whereby worker assemblies selected party members. In 1963, the ORI became the United Party of the Socialist Revolution (PURS). In October 1965, Fidel Castro convened the Central Committee (CC) of the Cuban Communist Party. Nearly 60 percent were military men. In comparison to the ORI National Directorate, the PSP share of the CC declined.[9] Moreover, Aníbal Escalante was not included; he had gone to Czechoslovakia on a prolonged vacation.

By 1965, the revolution thus had a vanguard organization. The party, however, did not legitimate the revolution; rather, Fidel Castro and the revolution bestowed legitimacy upon the PCC. The army, the militias, mass mobilizations, income redistribution, the social achievements, and national dignity embodied the power of revolution. Nonetheless, the organization of daily life under socialism often clashed with the popular sense of empowerment. The trade unions were a case in point.

Unions, Workers, and Conciencia

After November 1959, the revolutionary government and its supporters in the CTC leadership proceeded to establish tight control of the unions. Independent working-class activity—especially demands for higher wages and other economic benefits—was deemed contrary to the call for unity. Workers needed to develop *conciencia* of the new conditions: the imperative of pursuing policies to eliminate unemployment and satisfy the needs of the *clases populares* as a whole. Prior to the October 1960 nationalizations, the state had already demanded moderation from the working class; afterward, it completely disavowed the right or the need to strike. However, the new *conciencia* developed slowly.

In 1960, the establishment of technical advisory councils in nationalized enterprises was the first attempt to promote worker participation in management. The councils were to be "the experimental laboratory where the working class gains the experience for the future tasks of organically conducting national affairs."[10] Their objective was to educate workers about the production process and promote an awareness of the need to develop long-term strategies. Guevara hoped participation in the councils would lead workers to understand the necessity of "sacrificing an easy demand today to achieve a greater and more solid progress for the future."[11] Collective decision making was never their prerogative: the revolutionary government conferred exclusive power over enterprise matters to management. "Collective discussions, one-man decision-making and responsibility," Guevara contended.[12] Carlos Rafael Rodríguez seconded him: "We hear from many quarters the idea that workers should decide by majority vote. . . . Collective management is destructive. Administrators should have, have, and will have the last word."[13] Upon the demise of the councils in 1962, Guevara commented:

> The technical advisory councils constituted a first effort to establish meaningful links between workers and plant management. At that time, we manifested great prejudices about the ability of the working class to elect their membership adequately. . . . Mass participation in the elections was poor. The elections were bureaucratic.[14]

The Cuban leadership did not trust the working class to elect council members with the proper understanding of the new conditions. Economic demands had no place in the revolution. Cuban labor, however, had had a long history of unions, even while collaborating with the government, demanding wage increases and other benefits. The revolution decidedly revoked the tradition of militant reformism. The experience of the technical advisory councils reflected the problems of fashioning links between the global reality of revolution and the immediate circumstances of workers. Without doubt, the social revolution had granted the working class unprecedented gains. Undoubtedly too, more privileged workers stood less to gain—in some cases even to lose. At the end of 1961, however, more than 200,000 Cubans were still unemployed.[15] Job creation was the priority, and the revolutionary government would not contemplate sectorial demands.

In November 1961, a CTC congress—quite different from the one in 1959—was held. PSP union cadres and pro-unity July 26th Movement militants formed the new labor leadership. More than 9,000 delegates— the most ever in attendance at any CTC congress before or since—elected Lázaro Peña general secretary.[16] In contrast to 1959, delegates voted by a show of hands on a single slate of candidates. The unanimous acclamation of all proposals further undermined the credibility of the congress. Union conflicts in the recent past and continuing rumblings among workers surely indicated a diversity that found no expression at the congress. But socialism demanded a show of working-class support, and nationalism required ironclad unity in the revolution. Unions, moreover, would now be organized according to economic sectors: regardless of occupation or trade, all workers in an enterprise would belong to the same union. The organizational bases of the old union movement were dissolved.

The CTC congress spelled out the tasks for unions and workers: increasing production and productivity, saving raw materials, combating labor absenteeism, organizing volunteer work, safeguarding working conditions, preventing accidents, developing worker skills, promoting a new consciousness. The imperative of economic planning superseded the particular interests of workers. New wage scales and output quotas were instituted. Historical wages were, nonetheless, respected: salaries above the levels stipulated in the new wage scales would not be reduced. However, other historical benefits such as year-end bonuses and automatic sick-leave pay were rescinded. Strikes were decidedly ruled out.[17] Earlier Guevara had unequivocally asserted: "Cuban workers have to get used to living in a collectivistic regime and therefore cannot strike."[18]

Notwithstanding, difficulties beset the trade unions. The revolutionary government defined their role narrowly: acquaint workers with the point of view of the state, discipline workers for production, increase productivity, arbitrate between workers and managers.[19] Moreover, managers rather than union leaders often presided at worker assemblies. In union meetings, a "transparent wall" seemed to partition leaders from rank-and-file workers.[20] CTC General Secretary Lázaro Peña conceded that some union leaders "bureaucratically and with a truculent air" conveyed orders from above without adequate explanations, and workers were understandably resentful.[21] In 1963, Peña was himself the object of rank-and-file derision: construction workers facing a work stoppage heckled him when he suggested they accept substitute work at lower pay while equipment was being repaired and spare parts being purchased. Knowing full well an embargoed economy would not quickly be able to buy the needed parts, the workers rejected the proposition. Union leaders then proceeded to "educate" the workers and prepared more favorable conditions for discussing the need to prevent stoppages. A new assembly attended by fewer workers approved the suggestion of doing temporary work at lower pay.[22] Even so, some union leaders persisted in behavior reminiscent of the old Cuba. In 1963, the CTC National Council implored: "Our council calls on . . . all trade union leaders to rid themselves totally of the old concepts, limitations, worries, and language belonging to the capitalist past. Our tasks are new under the banner of the new society we are building."[23]

The revolutionary government did not deny the possibility of conflicts between the immediate interests of workers and the demands of a planned economy. Indeed, Guevara cautioned, "Should workers have to go on strike because the state assumes an intransigent and absurd position, it would be a signal that we have failed. . . . [I]t would be the beginning of the end for our popular government . . . but the state will ask for sacrifices from the working class."[24] He also noted: "The establishment of the socialist system does not eliminate contradictions, but rather modifies the way they are resolved."[25]

Between 1961 and 1964, grievance commissions settled conflicts between workers and managers. The commissions, however, backed workers against management too often. "The grievance commissions," Guevara complained, "will be able to accomplish a very useful task only provided they change their attitude. Production is the fundamental task."[26] Several months later, Fidel Castro conveyed his displeasure: "Many members of the grievance commissions seem to be on the side of absenteeism and vagrancy."[27] The Labor Ministry frequently reversed their decisions; in late 1964, the government abolished them and created work councils to enforce labor discipline more strictly. The newspaper *Revolución* outlined official expectations:

> The law will correct the mistakes and weaknesses of the former bodies of labor justice, but not by way of forgiveness. The new law will strengthen labor discipline and will increase production and productivity. . . . We have to recognize that . . . there are still undisciplined workers and for them we have to have discipline measures. . . . We still find workers who have not taken a revolutionary step and tend to discuss and protest any measure coming from the administration.[28]

Guevara recognized there were "deficiencies" in the relations of enterprise management, unions, and workers, and cautioned against "the old mentality, the boss mentality among plant managers, the exploited working class mentality among workers, fighting only for economic demands through the trade unions."[29] Similarly, Carlos Rafael Rodríguez urged union leaders to combat "the contempt against managers . . . which has flourished among broad strata of workers."[30]

Educating the working class was the avenue to overcoming the old mentalities. The revolution had instituted radical changes, and these created the groundings for a new *conciencia*. When these accomplishments were not persuasive enough on a daily basis, party and union cadres were supposed to convince workers—educate them—so that they understood the new conditions. The affirmation of national sovereignty, the promulgation of social justice, and the integrity of the revolutionary leadership—the dynamic of Fidel-*patria*-revolution—were the basis of the new politics. Within the logic of the revolution, there was no legitimate opposition to *la patria* and socialism, and opponents suffered repression or exile. The vanguard was charged with educating the workers who were "confused" and continued to insist on economic demands and antiadministration stance.

Fidel-*patria*-revolution, however, was less adaptable to the conduct of daily life. Although workers were said to be the owners of the means of production, the establishment of socialism caused many tensions among unions, enterprise management, and workers. *Conciencia* as owners did not come about easily and until it did, workers could not be fully trusted. Because the struggle for a new *conciencia* would go on for a long time, vanguard cadres were charged with orienting the rank and file and guaranteeing the loyalty of union leaders. Secret ballots and multiple candidacies were thus discarded. From the outset, the friction between a more general, future-oriented consciousness and one more specific and immediate confronted the revolution. The former was "correct" and congruent with national quests and socialist visions. Although often conflicting with long-term perspectives, the latter was more common. Cuban workers had forged a militant, economistic *conciencia* in their struggles against capitalism, and they did not quickly relinquish it.

The revolution, nonetheless, commanded the support of an overwhelming majority of workers—across racial, generational, sectorial, skill, and gender lines. The nationalization of the means of production had a profound impact on working-class consciousness. Workers identified the elimination of private property as a crucial factor in their heightened work commitment. "Never before has there been such fellowship between the workers and the administration and other Cubans," observed a cigar worker in 1962. More so than the citizens of the United States, the United Kingdom, Germany, Italy, and Mexico, Cubans indicated they expected fair treatment from their government. The affirmation of national sovereignty likewise contributed to working-class support for the revolution. "When a Cuban feels honor and pride in his heart for his nation," a brewery worker declared, "this means more than material benefits."[31]

The individual honor, national pride, and collective consciousness that ordinary Cubans manifested in the extraordinary times of Playa Girón, the Missile Crisis, the Literacy Campaign, and the first moments of the social revolution had inspired the advocates of moral incentives in the Great Debate. "One of our fundamental tasks is to find the way to perpetuate such heroic attitudes in everyday life," Guevara noted.[32] The radicals argued that new forms of economic organization and incentives would forge new patterns of social relations and *conciencia*. The moderates contended that material incentives were needed to buttress worker awareness. Alternate forms of economic organization and incentives would be feasible at a more advanced level of economic development. The Great Debate, however, did not focus separately on problems of political organization.

Only Foreign Trade Minister Alberto Mora, who was a proponent of self-financing and material incentives, alluded to the more complex setting in which a new *conciencia* developed. Mora emphasized the importance of politics:

> The only real way to assure the evolution of . . . socialist—and progressively communist—consciousness is to establish relationships of production within the framework of the new society's political organization (relationship of the individual to the State, the role of the Party, etc.) and ideological perception (of art, etc.). . . . Since the interrelationships in these areas are so strict, we are probably unable to assure the evolution of consciousness . . . by simply eliminating the desire for personal gain as a motive for social behavior. . . . Rather, we must at the same time assure the superstructure is so organized as to prevent the substitution of the money motive by the power motive.[33]

Weak institutions and organizations could not readily curb the arbitrariness of power so concentrated and unchecked. Moreover, corruption—one of the banes of the old Cuba—was on the increase. The masses wryly noted the development of *sociolismo*—obtaining preferential treatment for scarce material goods and other perquisites for and from *socios* (buddies). *La dolce vita* swept away not a few public officials, including union leaders.[34]

Nonetheless, the early 1960s began to address the challenges of consolidation. From the perspective of their fate after 1966, trade unions did not fare so badly. In 1975, then-CTC Foreign Relations Secretary Jesús Escandell observed:

> The period from 1961 to 1965 was a rich one for the trade union movement in Cuba. The CTC participated within its sphere. There were clashes with the administration, that is undeniable; some were justified, others were not. After the eleventh congress, which met in 1961, the trade unions participated in questions of salary, social security and other relevant tasks. The trade union movement participated decisively in the mobilization of volun-

teer workers who guaranteed the labor force for the sugar harvests during that period.[35]

Without question, trade unions were not independent and often clashed with workers who persisted in the old ways. Worker-union-management relations were tracked on an upward spiral: union leaders and managers ultimately depended on those above them for their jobs. Nonetheless, unions existed and functioned "within their sphere." During the late 1960s, when trade unions "withered away," state and party relations with rank-and-file workers would be even more precarious. An organization with limited autonomy was certainly better than no organization at all.

The Federation of Cuban Women

Unlike the CTC, the Federation of Cuban Women was relatively free of conflicts. Cuban leaders did not consider gender to be central to the revolution, as class was. The revolutionary government had to maneuver to gain control of the long-standing CTC. Born with and for the revolution, the FMC gave many women their first opportunity to have a life outside the home. Women constituted a reservoir of support for the revolution, and the FMC readily tapped it. While workers endorsed the revolution, their everyday *conciencia* formed in union struggles before 1959 often clashed with the prerequisites of central planning. Starting *tabula rasa* proved to be less problematic: the revolution would forge the *conciencia* of women.

In 1962, over 4,000 delegates attended the FMC congress. At the time, FMC membership totaled 376,000. The congress ratified the purpose of mobilizing women on behalf of the revolution and took pride in the record of the organization. More than 19,000 women who had been household servants had graduated from special schools and were now otherwise gainfully employed. The seamstress programs had trained 7,400 rural women in the use of sewing machines and now they were instructors to 29,000 young peasant women. The FMC gave first-aid training to nearly 11,000 women, mobilized 62,000 for volunteer work, and managed over one hundred day-care centers. Having participated in the Literacy Campaign, the federation was now helping the Education Ministry to administer scholarship programs for 70,000 students.[36] The FMC likewise supported the Public Health Ministry in the promotion of personal hygiene and pre- and postnatal care, especially in rural areas. At the congress, FMC president Vilma Espín noted: "The ideal new woman is a healthy woman, mother of the future generations who will grow up under communism."[37]

The FMC also established chapters in factories. In 1963, the CTC and the federation cosponsored the National Conference on Women

Workers, which emphasized the strides women had made in certain sectors. More than 800 women were union leaders in the food industry, where more than 10,000 women worked. Women held administrative positions in 4,500 commercial enterprises. More than 900 former household servants were now employed as bank clerks. There were more than 1,100 women union leaders and more than 1,200 women workers with vanguard status in public administration. There were 863 nurses, 2,377 aides, 1,250 student nurses, and 2,809 health activists in the public health sector. About 83,000 women were employed in agriculture.[38] Because of differences in categories, pre-1959 comparisons are not exact but the number of women in most of these sectors appears to have increased.[39]

Nonetheless, the percentage of women working declined during the early 1960s. In 1964, there were 282,069 working women (11.3 percent), barely surpassing the number working during the 1950s.[40] The numbers, however, do not reveal the magnitude of the changes. Professional and executive women had accounted for nearly 20 percent of the female labor force in 1953. By 1963–1964, many of them were probably in the United States. The revolution, moreover, virtually eliminated household service, a category that had employed over 25 percent of women workers in 1953. A sizeable proportion of the 282,069 working women in 1964 were, therefore, new entrants or were employed in jobs different from those that they had held during the 1950s. The expansion of employment during the early 1960s benefited men more than women.

Although not as seriously as the CTC, the FMC also experienced problems as a mass organization. In 1963, the federation acknowledged that cadres were mechanically carrying out their tasks and were generally inattentive to members of the rank and file, who were themselves becoming apathetic. Cadres too frequently followed directions from above without creative adaptation to their specific chapters, and communicated the directions to their members as orders. The FMC pointedly noted:

> It is not enough that directions be issued, a newsletter printed and sent to the provincial chapters. We have to attend to how these orientations are communicated to the base of our organization, how the masses interpret them, what are their opinions. Their judgments and opinions enrich and advance our revolutionary work.[41]

There was thus a general loss of popular élan. Incipient institutionalization was not incorporating the vitality of Fidel-*patria*-revolution. New forms of alienation were becoming evident as the spirit of 1959, Playa Girón, the Literacy Campaign, and the Missile Crisis receded, and the reality of socialism—the vanguard party, mass organizations, and a disappointing economic performance—was looming larger in the popular *conciencia*. Radical change was in the offing.

The Origins of the Radical Experiment

Under socialism, daily life did not generate the energy and enthusiasm manifested in times of victory, aggression, or crisis. The revolution brought down the old bastions of privilege, improved living standards, and empowered ordinary Cubans to take up arms for national defense. That was the substance of democracy. How to translate substance into practice was another matter. Cuba adopted then-extant socialist models of politics and economics because they complemented national needs and were the only alternative to capitalism and representative democracy. The experience of 1961–1965, however, fell short of expectations.

Central planning did not prove to be the highly touted panacea to underdevelopment. By 1963, rapid import-substitution industrialization had failed. With its emphasis on sugar and agriculture, the new development strategy required the mobilization of labor to rural areas and the promotion of high rates of capital accumulation. Between 1962 and 1965, however, rural labor declined from 38 percent to 32 percent of the labor force.[42] Poor economic performance, moreover, dampened expectations for rapid increases in consumption. Initial improvements in standards of living would not be easily replicated. In 1962, rationing of food, clothing, and most consumer items was established. Although *la libreta* (the ration book) guaranteed equality in the distribution of basic goods, it did so in austerity, not in the prosperity first envisioned. With the establishment of rationing, the state was implicitly acknowledging that central planning generated its own irrationality and chaos. Further, the priority given to investment and capital accumulation limited the resources available for immediate consumption. Thus, the shift toward agriculture necessitated a policy to address the rural labor shortage temporarily because mechanization was the long-term solution. Although mounting shortages and investment priorities preempted that mobilization on the basis of wage differentials and other material rewards, the revolution was also reluctant to breach its commitment to social justice.

The organization of politics had likewise highlighted that one-party dynamics could not be allowed to unfold like a new "invisible hand" without compromising the revolution. Sectarianism in the party underscored the imperative of new recruitment methods to secure "contact with the masses." A vanguard party deserved its name only if it maintained an organic relationship with a people in revolution. Party militants—not party structures—were the crucial link in that relationship. "Cadres are the masterpieces, the ideological motor . . . creative individuals. . . . [T]hey help in the development of the masses and in keeping the leadership abreast," Guevara forcefully asserted.[43] People—not a "correct" party line—were the key to the vanguard. Ideological purity was not the grist of party militants; the crucible of day-to-day struggle was. The ORI crisis reinforced in the revolutionary leadership a mistrust of the formalization of political power. A party could be under-written by the "letter" of the Marxist-Leninist manuals—the new catechisms—and extinguish the "spirit" of Cuban reality.[44]

The Cuban "spirit" was also dulled by another consequence of so-cialism: an expanding state bureaucracy. Ministries, agencies, institutes, committees, and meetings proliferated. In 1961, the JUCEI—local com-missions of coordination, implementation, and inspection—were estab-lished to regulate and supervise government offices. Party-appointed JUCEI delegates were charged with mediating between the state, the mass organizations, and the public. The commissions, however, were quite hierarchical and were staffed by full-time personnel. Bureaucratic expansion was dimming the prospects of popular control. Moreover, unlike the vanguard and the workers with *conciencia*, functionaries were more interested in pushing the "letter" of papers than in advancing the "spirit" of national quests and socialist visions. In 1964, the party launched a campaign against bureaucratism. Fidel explained:

> It is necessary that we avoid the inception of a parasitic class living at the expense of productive work. . . . If we fill buildings with employees, we will be more expensive to the country than the old politicians were. . . . We have accomplished nothing if we previously worked for the capitalist and now we work for another type of person who is not a capitalist, but who consumes much and produces nothing. . . . [T]he standards of living of the people cannot be raised while what one produces must be divided by three.[45]

Thus, Cuban leaders believed that unchecked bureaucratic expansion limited economic performance, and that curtailing it would increase productivity. The government mobilized former bureaucrats to work in agriculture; in rural Cuba, staid functionaries would confront the reality of revolution and in the process develop a new *conciencia*.

International conditions also contributed to the origins of the radical experiment. When import-substitution industrialization failed, one of the reasons for returning to sugar was the need to generate hard cur-rency. After 1970, when the strategy would supposedly tender its first benefits, an expanded sugar sector would grant Cuba the resources to diversify the economy, eliminate trade deficits, and lessen the new de-pendence. Moreover, the Soviet Union seemed less powerful, reliable, and respectful of Cuban sovereignty than Cuban leaders had first thought would be the case. Indeed, the Missile Crisis revealed that the Soviet Union was no match for the United States. In addition, the Soviets had agreed to on-site inspection of Cuban territory without prior consul-tation with the Cuban leadership. When the United States began over-whelming air raids against Vietnam in 1965 and the Soviet Union failed to respond forcefully, Cuban leaders became convinced that their na-tional defense ultimately depended on Cuban resources. Thus, promoting revolutionary change in Latin America became more critical. With allies in the Western Hemisphere and more economic resources, Cuba would lessen its dependence on the Soviet Union. The revolution would then be able to mark its own road to socialism without concessions to orthodoxy at home or abroad.

The Parallel Construction of Communism and Socialism

When Fidel Castro introduced the Central Committee of the Communist party in October 1965, the foundations for the radical experiment were in place. The old communists did not have the preeminence they had had in the ORI. Although prominent PSP leaders like Blas Roca and Carlos Rafael Rodríguez were members of the Secretariat, the Politburo did not include a single old communist. The Central Committee incorporated no one who had been associated with Aníbal Escalante. The PCC was firmly in the hands of the new communists, especially those with military or Sierra Maestra credentials. The party scheduled its first congress for 1967.

The PCC rejected long-established dogmas of international communism. Cuban communists declined to recognize the leadership of the Soviet Union and reserved the right to formulate their own foreign policy. "We will never ask anyone's permission to go anywhere," Fidel declared.[46] After a brief courting of Latin American Communist parties in 1963–1964, the Cuban leadership turned to other left organizations and movements. The banner of Marxism-Leninism belonged to those who actually made revolutions. "The duty of every revolutionary is to make revolution," proclaimed Fidel in the Second Declaration of Havana in 1962. Ernesto Guevara was not among the members of the new Central Committee; he had gone to other lands to continue the "struggle against imperialism."[47] Ché embodied the international and domestic orientations of the radical experiment.

The revolution aimed to recover its initial impetus that the early efforts at party building, mass mobilization, and economic planning had partially stymied. The tasks now were to struggle against bureaucracy, thwart petit-bourgeois values, improve the efficacy of state administration, attain a 10-million-ton harvest in 1970, and organize *poder local* (local power). Under charismatic guidance and with the masses as a "constant fiscal agent," the revolution would "develop [its] own revolutionary path . . . in creative ways, taking advantage of our people's rich imagination and great intelligence . . . with great self-confidence." Superseding the JUCEI, *poder local*—"a school of government"—was an attempt to incorporate ordinary citizens into state administration "in a society where the masses exercise maximum, total participation." Secret ballots and direct elections would not, however, be the means to select local power delegates. The PCC would instead exercise close scrutiny over the selection process. A congress of *poder local* was also scheduled for 1967.[48]

The "organic consolidation" of the PCC was at the center of institutional renovation and state reorganization.[49] Membership had grown more than threefold since 1962, and cadres had stronger links with *las masas* because of the new recruitment method. The party was the sole and purest expression of the popular will and had the exclusive right to educate the people.[50] The Cuban revolutionary experience underscored the role of dedicated individuals and the primacy of visionary leadership. The sectarianism crisis had, moreover, reinforced the importance of human agency: without cadres, organizational structures and "correct" directives were meaningless.[51] Noted Fidel:

> Undoubtedly, voices will be raised to appeal to people's selfishness. But those of us who consider ourselves revolutionaries will never cease to struggle against individualistic tendencies and will always appeal to the generosity and solidarity of our people.[52]

The antibureaucratic campaign was central to the radical experiment. Because the bureaucracy had more control over the means of production than the people, the revolution faced the danger of "a special stratum of citizens" who "can convert bureaucratic positions into places of complacency, stagnation, and privilege."[53] Bureaucrats paid more attention to their superiors in the central ministries than to the people and local enterprises. Removed from the base where the revolution was taking place, they obstructed the effective administration of the economy. Reducing the numbers of people employed in nonproductive activities was therefore imperative. Many bureaucrats were transferred to more productive tasks, particularly in the countryside. Between 1965 and 1967, about 1 percent of the labor force was "debureaucratized"; nearly half of those laid off lived in Havana.[54] Combating the bureaucratic malaise also demanded massive investments in education. Cadres and the people could not properly supervise the bureaucracy if they were not educated.[55]

The campaign, however, was against bureaucratism, not planning. In 1965, the party consolidated the management of the economy in JUCEPLAN and the National Bank. Noted President Osvaldo Dorticós: "The theories of socialist construction and socialist planning are an incipient science. . . . [T]he Cuban Revolution . . . should be capable of contributing to their scientific development."[56] A new style of administration—simple, agile, rational—was needed: about 1,500 positions were eliminated.[57] The consequences of the reorganization, however, were not salutary: JUCEPLAN and the National Bank saw their ability to administer and plan severely constrained. Party directives abolished enterprise payments and receipts, taxes, interests, cost accounting, and other economic controls. Between 1967 and 1970, budgets and plans were discarded in favor of *fidelista*-improvised miniplans. Wages were divorced from productivity and quality of output, payment for overtime eliminated. The revolution intended to "equalize incomes from the bottom up, for all workers, regardless of the type of work they do."[58]

Like Guevara and others in the Great Debate, Cuban leaders now advocated moral incentives and the curtailment of market relations. Guevara, however, had insisted on a system of planning: strict budgetary controls, wage scales linked to output quotas, and a mixture of moral and material incentives. Improvisation, extreme egalitarianism, economic disorganization, and disregard for administrative procedures were contrary to budgetary planning.[59] Nonetheless, the moderates in the Great Debate had admonished that overcentralization could result in the breakdown of planning. Fidel Castro aptly summed up the spirit of the radical experiment:

> What is important is that we have a surplus of production and not of papers, even though there might not be a single paper tallying those products. . . . When we have surplus production, our problems will be of another nature. What interests us, in any event, is to record our surplus and not to accumulate files on the deficit.[60]

Thus, the antibureaucratic campaign undermined planning and contributed to the economic chaos of the late 1960s.

The policies of the radical experiment shared with the radicals in the Great Debate a paramount concern with the development of *conciencia*. Fidel sounded a favorite Guevara theme when he said:

> We will not create a socialist consciousness, much less a communist consciousness, with the mentality of shopkeepers. We will not create a socialist consciousness with a dollar sign in the minds and hearts . . . of our people. . . . We will not reach communism by using a capitalist road. Using capitalist methods no one will ever reach communism.[61]

Socialization of the means of production was not sufficient to secure the transition to communism. A new culture based on collective needs was also imperative. Thus, the radical experiment rejected the familiar paths in the Soviet Union and Eastern Europe and exalted Cuban "means, procedures, and methods" to build socialism and communism.[62]

Neither Guevara during the early 1960s nor Fidel during the late 1960s, however, envisioned a link between *conciencia* and formal democracy. Both understood popular participation in terms of widespread and enthusiastic involvement in the revolution under charismatic guidance and vanguard party leadership. Guevara, however, did dwell on the problems of working-class organization under socialism. By the mid-1960s, though, the orthodox model of unions—like those of the vanguard party and economic planning—had fallen into disrepute. After 1966, vanguardism overwhelmed the trade union movement, which in effect "withered away." Unions as a mass organization ceased to exist.

The Withering Away of Trade Unions

Thse CTC congress of 1966 met amid an unfolding radicalism. The struggle against bureaucracy, the strategy of agro-centered development, and the centrality of *conciencia* dominated the gathering. The congress attributed the difficulties of the early 1960s to bureaucratization and professionalization. Union leaders had become professionals and more responsive to directives from above than to the people. Too many leaders and directives "hindered the correct political orientation of the union movement" and "spawned a strong bureaucratic machinery," Trade unions were gripped by "'meeting-itis' and 'coordinating-itis' . . . a rampant pass-the-buck attitude" that thwarted "workers' enthusiasm and active participation."[63] Nearly 75 percent of local union officers were turned out in the precongress elections. CTC full-time personnel were reduced 53 percent; the number of cadres declined from 2,227 to 968.[64] New local leadership and a streamlined national administration were the key to mobilize the rank and file in support of the new policies. Agricultural development was to be the center of the union movement.

Unlike the congress of 1961, the 1966 congress did not provide a framework for the role of unions under socialism. Issues such as worker input in enterprise administration and the defense of workers' rights were not on the agenda. Questions of output quotas, wages, and work organization were only generally addressed. However, the congress did insist on the character of unions as a mass organization distinct from the party and the state administration and made a commitment to developing more responsive union structures. The CTC also implied that the union movement—not the Labor Ministry—should have jurisdiction over the work councils and the enforcement of labor discipline.[65] The old PSP labor chieftain Lázaro Peña left his post at the helm of the CTC and assumed responsibility for a new commission on labor matters adjunct to the Central Committee.

The CTC congress was the forum for a fuller exposition on the importance of moral incentives. Miguel Martín, the new CTC general secretary, focused on the "new man" as crucial in the transition to communism. Living standards would improve primarily by the distribution of collective goods such as education and public health.[66] Individual incentives undermined *conciencia* and defeated the purpose of revolution. The most precious resource was the support of *el pueblo cubano;* their will, energy, and passion had to be directed toward economic development. If the strategy succeeded without resorting to material incentives, popular *conciencia* would be enhanced. The CTC congress also served as platform for Fidel Castro to reiterate the international resonance of the radical experiment:

Under socialism products should not be sold according to production costs, but according to their social function. . . . We are capable of leaving the manuals aside, we dare to exercise the right to think. . . . We do not belong to a sect, we do not belong to an international masonry, we do not belong to a church. We are heretics, well we are heretics: let them call us the heretics.[67]

During the late 1960s, emphasis on moral incentives and the reduction of union cadres contributed to the demise of the labor movement. Union leaders tended to be party members and never lasted too long on their jobs because they were transferred to other positions. That unions provided cadres for other institutions was an indication that Cuban socialism was drawing cadres from the working class. The turnover of local union leaders, however, also underscored other realities. In practice, there were no institutional boundaries between the party, management, and the unions. Moreover, unions did not seem to be important. How could mundane concerns over work organization and the rights of workers compare with the mission of finding the Cuban road to socialism and communism? In fact, local unions became organizations of vanguard workers who constituted no more than 20 percent of the labor force. Ordinary workers with the wrong *conciencia* had no organization.

Undoubtedly, moral incentives moved vanguard workers to heroism, sacrifice, and dedication. Vanguard workers manifested a "class-for-itself consciousness" and a "social perspective."[68] They understood the need to subsume "particular interests" to the historical quests of the revolution. *Conciencia*—not time cards, job dismissals, wage cuts, and other economic sanctions—motivated their labor discipline.[69] Average workers, however, were progressively demoralized. Although austerity and egalitarianism rendered their wages meaningless, national quests and socialist visions failed to motivate them. On average, about fifteen days of work provided the money needed to purchase rationed goods. In 1969–1970, absenteeism reached 20 percent of the labor force; some areas in Oriente province registered an astounding 50 percent rate. A 1968 survey of two hundred enterprises indicated that 25 to 50 percent of the workday was wasted. Volunteer work lost the original purpose of mobilizing workers to do needed tasks beyond their regular work. Instead, the party called on workers to put in extra hours to complete what should have been accomplished during the normal day. The process of labor justice weakened considerably as grievances declined 50 percent. Ordinary workers had no recourse to express their dissatisfaction except absenting themselves from work.

Nonetheless, the newspaper *Granma* asserted that "our labor movement is today at a superior stage of development" and continued the appeal to "revolutionary *conciencia*, our sense of honor."[70] The "strength" and "sweat" of the people were "capable of creating imcomparably superior riches" to the "adding and subtracting" of the "pure economists."[71] The revolution itself motivated workers: immediate concerns were secondary. "It is not a question of discussing all administrative decisions with the workers," Politburo member Armando Hart noted, "but of obtaining their enthusiasm to support the principal measures of the administration."[72] Some Cuban leaders, however, manifested another understanding of the situation of the trade unions. In 1969, Carlos Rafael Rodríguez observed:

> The unions are transmission belts for Party directives to the workers, but they have insufficiently represented workers to the Party or the Revolutionary Government. . . . They cannot be mere instruments of the Party without losing their purpose. Administrators after all can also be *hijos de puta* (sons of bitches), and if they are, workers have to be able to throw them out and for that matter, do the same with any bureaucrat.[73]

By comparisons to the CTC, the FMC fared rather well as a mass organization during the radical experiment. In 1970, membership totaled 1.3 million.[74] The FMC continued mobilizing women for numerous tasks. By 1968, 55,000 young peasant women had learned seamstress skills. Over 700,000 women had received instruction in health care and personal hygiene.[75] In 1970, the FMC operated 433 daycare centers with a total capacity for more than 47,000 children.[76] When the party disbanded FMC factory committees, the CTC assumed responsibility for women workers. With the demise of unions, the needs of working women also went unattended. Still, the mobilization of female labor in urban Cuba to make up for male labor mobilized for agriculture became one of the FMC's central charges.

Although women increased their share of the labor force, they experienced a high turnover. In 1970, there were 482,257 women in the labor force (18.3 percent).[77] However, one in four women who entered the labor force dropped out within a year. Nonetheless, the structure of female employment continued to change. Two in five women were employed in social services, one in five in industry, nearly one in four in commerce, and about one in twelve in agriculture.[78] Thus, the trends of the early 1960s continued throughout the decade. In 1970, there were 23,064 cadres in political and mass organizations, 6,475 (28.1 percent) of whom were women.[79] During the late 1960s, the proportion of women in leadership positions across Cuban society was around 9 percent.[80] The federation, however, did not express a feminist understanding of gender inequality.[81]

The Politics of Mobilization

During the late 1960s, the revolution aimed to imprint a Cuban face on contemporary socialism. Cuban leaders wanted to develop organizational models that better suited Cuban culture and history. They looked to their own experience in the struggle against Batista and in the radicalization of the revolution. Visionary leadership, steely determination, and popular support had been the keys to their success. The radical experiment sought to capitalize on the most important resource: the will, energy, and passion of *el pueblo cubano*. Producing 10 million tons of sugar in 1970 was more than an economic goal: "a point of honor for this revolution, a yardstick by which to judge the capability of the Revolution." Failure would mean the Cuban people would have to "draw in our horns, be more calm, more docile, more submissive—in short, cease being revolutionaries."[82] The harvest, however, produced 8.5 million tons—the largest ever, but well short of the honorable 10 million. The challenge to orthodoxy had failed.

The radical experiment floundered almost from the start. Rallying the Cuban people around national quests and socialist visions took precedence over institutions and organizations. The announced PCC and *poder local* congresses never took place. Vanguard workers who were a minority loomed larger than ordinary workers. In 1968, Hector Ramos Latour was appointed CTC general secretary without the formality of an election at a national congress. After 1966, local union elections were suspended even though CTC bylaws required them every two years. Grievance procedures were virtually eliminated. The FMC lost its factory chapters in detriment to the interests of women workers. The PCC itself essentially stagnated. Until 1969, membership remained at about 50,000. By 1970, however, party members had more than doubled to more than 100,000, about 1 percent of the population. Rank-and-file workers, however, were not the basis of the new growth: the central ministries, the armed forces, and the Interior Ministry were.[83]

In 1967, the radical experiment was dealt a serious blow. On October 7, Ernesto Guevara died in Bolivia. The Andes would not become the Sierra Maestra of Latin America. Because Cuba would not soon benefit from the support of other revolutions, success at home was the only avenue left for the radical experiment. In January 1968, there was a breach in the apparent consensus among Cuban leaders on the radical policies. A group of old communists had formed a "microfaction" within the party and were consulting Soviet embassy officials on changing the foreign and domestic policies of the revolution. They were expelled for agreeing with the "pseudorevolutionaries" that Cuban policies were "adventurist" and "unrealistic."[84] Within the logic of Fidel-*patria*-revolution, dissent was intolerable. The microfaction members, moreover, had behaved like the old *políticos* who had always consulted the U.S. embassy. The Soviet Union had earlier announced a delay in petroleum shipments, and Cuba was once again reminded of its dependence.

Nonetheless, the radical experiment continued. In March and April 1968, the state nationalized more than 58,000 small business establish-

ments and banned self-employment. Ironically, more than half of the
nationalized businesses had started operations after 1961. Before the
nationalizations, the state had controlled less than 25 percent of retail
trade and sales. In Las Villas and Matanzas, an overwhelming majority
of these transactions were in private hands. Private ownership had even
extended to small industrial production. In 1967, for example, the state
purchased agricultural tools for 628,000 pesos from a private manufac-
turer who employed eighty-nine workers. These small entrepreneurs,
moreover, constituted a focus of opposition to the revolution, supplied
the black market, and reinforced the values of capitalism.[85]

On August 23, 1968, Soviet and Warsaw Pact forces invaded
Czechoslovakia. Under the leadership of Alexander Dubcek, the Czech
Communist Party had initiated a process of socialist renovation through
market-oriented reforms and political liberalization. Czechoslovakia
claimed the right to bestow a human face on contemporary socialism.
The old Soviet Union, however, could not tolerate diversity and independence in Eastern Europe. Cuba extended qualified support to the
invasion. On the surface, Cuban support was surprising. Relations between Cuba and the Soviet Union were tense. Czechoslovakia was a
small country whose sovereignty had been preempted by a large and
powerful neighbor. Fidel deplored that Czech communists had so devi-
ated from socialism that such an infringement became necessary. The
Prague Spring had promoted a vision of socialism that ran counter to
that of the radical experiment. The invasion provided Cuban leaders
with the opportunity to assert their opposition to market socialism and
extend a conciliatory gesture to the Soviet Union.[86]

The year 1968 also marked the centennial of the nineteenth-century
wars against Spain. "Our revolution is one revolution," asserted Fidel,
"and it began on October 10, 1868."[87] After 1959, socialism had realized
the nineteenth-century quest for national independence and social justice that the revolution of 1933 had rekindled, the *políticos* had betrayed,
and *el pueblo cubano* had never forgotten. The assault on Moncada Barracks redeemed the legacy of José Martí, and the revolution had subsequently rendered Cuba a sovereign nation. Had the *fidelistas* followed
the manuals, the revolution would not have come to power. In 1968,
Cuba was daring to leave its mark on contemporary socialism, and
orthodoxy would once again be proven wrong.

The 1970 Watershed

The radical policies of the late 1960s revealed a number of dramatic
pitfalls. As capitalist incentives and structures receded and socialist sub-
stitutes failed to emerge effectively, militarization filled the void. Mass
mobilizations, however voluntary for thousands of Cubans, passed under a regimen of military discipline. *Conciencia* did not inspire people to

work for the collective well-being. Instead, they chose to swell the ranks of absenteeism, waste the working day, and start their own small businesses. Many young Cubans were seduced by the counterculture of the 1960s in the United States. The hippies—not the work ethic—were the "American way." Vandalism and theft flourished in Havana.[88]

Daily life assumed increasingly baneful qualities. Bars were closed and vacations postponed until after the heroic, decisive harvest. Although petty retail trade offended revolutionary morality and undermined *conciencia*, it also provided goods and services that the state sector could not. Corner stands offering coffee and a quick snack disappeared. Family stores peddling toiletries, household goods, and other items vanished. Dry cleaning, appliance repairs, plumbing, and a host of other services became unavailable or were mishandled by the state. Consumption was at least 5 percent below 1962 levels.[89] Indeed, the economy nearly collapsed. Some Cuban leaders had earlier addressed the causes of the mounting chaos and looked beyond the decisive harvest. Osvaldo Dorticós, for example, had expressed concern about the "abuse" and "deceit" of overtime, the breakdown of controls over production and labor norms, and the promotion of revolutionaries lacking qualifications to management positions.[90] In 1969, Raúl Castro and Carlos Rafael Rodríguez had summoned separate groups of economists to make plans for reorganizing the economy after 1970.[91]

On July 26, 1970, Fidel Castro confronted the crisis: "Our enemies say we are faced with difficulties, and in fact our enemies are right."[92] Yet there was no alternative to the revolution and socialism. However discontent, the people were not about to choose counterrevolution and capitalism. Fidel accepted responsibility for the debacle and implied he would resign if the people willed it. But, he said, there were no magic solutions. Even though the learning process of the leaders had been costly, now that they had learned, changing them would not resolve the crisis. Thus, there was also no alternative to Fidel. Chastened by their mistakes, the same leaders would give the revolution new directions. Fidel noted that low levels of education among cadres was one of the foremost problems: nearly 80 percent of the party membership had less than a sixth-grade education.[93] Resorting to the people—their *conciencia*, their decisiveness, their commitment—was the only alternative, Fidel emphasized. Factories needed more collective forms of management, and the input of rank-and-file workers was imperative.[94]

Indeed, workers had sent the leadership a powerful message: absenteeism was so extensive that they had, in a sense, staged an uncoordinated strike.[95] Consequently, the revitalization of the labor movement was an immediate priority. There would be "absolutely free" elections to reconstitute local trade unions.[96] Administrative fiat would now give way to democratic procedures. The CTC would be strengthened and join the FMC and the Committees for the Defense of the Revolution as powerful mass organizations.[97] "Without the masses, socialism loses the battle," affirmed Fidel, "it becomes bureaucratized . . . and has to resort to capitalist methods."[98] Bureaucratic rule was not, however, entirely dependent on the size of the bureaucracy. Reducing the number of bureaucrats had not changed the nature of the bureaucracy. The concentration of power in public officials who were not answerable to the people was the essence of bureaucratization.[99] Castro even suggested an inverse relationship between the vanguard and the rank and file:

> We speak of infusing a proletarian spirit, creating *conciencia*. It is a lie. We are today in a situation in which we have to go to the factories where the working class is . . . to learn from the workers. We do not take *conciencia* to the workers.[100]

Nonetheless, Fidel argued that the party retained the right to rule because its cadres were not corrupt.

The unmet 10-million-ton harvest indeed represented more than a failed economic goal. The revolution had miscarried the attempt to generate economic and political resources to imprint a Cuban face on contemporary socialism. And now, institutionalization could no longer be postponed. There was no alternative but to turn to the models of the Soviet Union and Eastern Europe. There was also no option but to accept the new dependence and live with its consequences. Precedence had to be given to civilian pursuits and economic development. By the end of the 1960s, the Cuban people had manifested a *"resignación de apoyo"* (resigned support).[101] Indeed, the year 1970 poignantly marked the end of the revolution. The social bases of political power had been transformed, and the institutionalization that was about to begin would impart a more settled dynamic on Cuban society. Charismatic authority, vanguard politics, and mass mobilizations would subsequently acquire a new context. The reality of socialism would slowly gain ascendance over the effervescence of revolution.

CHAPTER 6

Politics and Society, 1971–1986

Even without representative institutions, our revolutionary state is and always was democratic. A state like ours which represents the interest of the working class, no matter what its form and structure, is more democratic than any other state in history.

Raúl Castro
August 1974

The Cuban Revolution failed to take advantage of the rich experience of other peoples who had undertaken the construction of socialism before we had. Had we been humbler, had we not over-estimated ourselves, we would have been able to understand that revolutionary theory was not sufficiently developed in our country . . . to make any really significant contribution to the theory and practice of socialist construction. . . . It was not a matter of mere imitation, but of the correct application of many useful experiences.

Fidel Castro
December 1975

We have to avoid compromising our communist *conciencia* with socialist formulas. . . . It is good that people work harder because they earn more. . . . We produce more, but it is not a communist attitude. . . . The development of communist society must go hand in hand with increasing our wealth . . . otherwise it may be that our wealth increases and our *conciencias* are weakened.

Fidel Castro
April 1982

Institutionalization imprinted Cuban socialism with a familiar face. The Communist party expanded its membership, broadened its leadership, and established a formal apparatus. Between 1975 and 1986, three congresses were held. The Central Organization of Cuban Trade Unions and the Federation of Cuban Women likewise held congresses on a regular basis and fulfilled their role as "transmission belts" between the PCC and the people. Vanguard party politics allowed for the limited expression of sectorial interests. In 1976, the party adopted an economic management and planning system of relative decentralization and material incentives. In Popular Power assemblies, the citizenry had the means to exercise a modicum of control and supervision over local matters. The delegates, moreover, were elected by means of secret ballots and multiple candidacies. In a 1976 referendum, 97 percent of the electorate approved a new constitution. Under Communist party control, voting secured a place within Cuban socialism.

The Cuban government established closer ties with the Soviet Union and Eastern Europe. Cuba joined the Council for Mutual Economic Assistance and obtained new credits, debt postponement, and preferential terms of trade. New realities in Latin America and Africa provoked transformations in Cuban foreign policy. State-to-state relations—not guerrilla movements—characterized emerging links with Latin America. With Soviet support, professional military expeditions marked Cuban internationalism in Africa. Novel means guided the pursuit of original purposes. Activism in world affairs gained the Cuban government a measure of security and independence. For a time, even the outlook for relations with the United States seemed optimistic.

Revolution and Institutionalization

The outcome of the radical experiment had underscored the importance of institutions. Without them, there had been no check on public officials, the economy had gone into chaos, and workers had become demoralized. Mobilization had been no substitute for participation; cadres with *conciencia*, no surrogate for organization. Drawing upon the legitimacy of Fidel Castro and the social revolution, the Communist party pursued a process of institutionalization. As it had during the early 1960s but more thoroughly and systematically, Cuba turned to the Soviet Union for models of economic and political organization. Cuba was not the Soviet Union, however.[1] The Cuban Revolution had come to power only in 1959, and the government still commanded substantial popular support. Although without quite the same fervor with which he had promoted the radical experiment, Fidel Castro embraced the institutionalization. The dynamic of Fidel-*patria*-revolution nonetheless continued at the center of Cuban politics.

The Cuban leadership never considered that pluralism and divergence would mark the new directions. On the contrary, the purpose of the institutionalization was to confirm socialism and the leading role of the PCC. The citizenry did not have the right to opt out of socialism or to challenge the party and the leadership. The social revolution itself continued to be the fount of legitimacy. Liberation from a past of national subordination and social inequity validated the present. One of the principal charges of institutionalization was the differentiation of political leadership, administrative responsibility, and popular involvement: the PCC was to rule, the state to administer, the mass organizations to maintain "contact with the masses." The armed forces played a crucial role in the initial period of institutionalization. As the only institution to survive the 1960s virtually intact, the military offered "civic soldiers," who assumed numerous and varied assignments in civilian life.[2] A constitution sanctioned the new order. During the 1970s, Cuba assembled a socialist polity.

The Organs of Popular Power (OPP) embodied the process of institutionalization. After a 1974 pilot in Matanzas, the OPP were structured nationwide in 1976. Popular Power was similar to the old *poder local* that the 1960s had never quite instituted. Municipal, provincial, and national assemblies were constituted to supervise the state administration. Thus, supervision of schools, clinics, grocery stores, garbage collection, maintenance shops, movie theaters, and small local industries was transferred to municipal assemblies. By the early 1980s, more than a third of the national economy was under local Popular Power intendance.[3] Between 1977 and 1983, local industries under OPP supervision tripled their output in value.[4]

Unlike *poder local,* the party did not appoint delegates to municipal assemblies. Although no campaigning was allowed, citizens elected their local delegates through secret ballots and multiple candidacies every two and a half years. Municipal assemblies elected the membership of provincial assemblies and the latter elected the delegates to the National Assembly. At least 55 percent of the delegates to the National Assembly were supposed to have been elected in the municipalities. The other 45 percent were selected from a list of candidates proposed by the party leadership. One of the duties of the National Assembly was confirmation of members of the Council of State and Council of Ministers submitted by the Politburo. The councils and Popular Power were under tight party control. Nearly 50 percent of the Council of State and about 25 percent of the Council of Ministers were Politburo members. Over 80 percent of the members of the Council of State and 65 percent of the members of the Council of Ministers were members of the Central Committee. More than 90 percent of National Assembly delegates were members of the Communist party. About 75 percent of all local delegates were PCC or Communist Youth militants.[5] The separation of political leadership, popular supervision, and state administration was not achieved.

Municipal assemblies and local delegates constituted the most important link between the state and the citizenry. Three times a year, delegates engaged their constituencies in assemblies of *rendición de cuentas* (rendering of accounts). Two years after the establishment of Popular Power, the National Assembly heard a report that noted the local meetings had become formalistic "to such an extreme that at times delegates prepare their presentations beforehand and do not fully express their opinions."[6] Too often delegates used the same rationale as ministers and managers to justify problems. They were not doing their job, and attendance to *rendición de cuentas* had consequently declined. The National Assembly did not receive the report well. National delegates argued that local delegates were being held responsible for problems originating with state functionaries. Ministries, government agencies, and enterprise administrations refused to recognize the authority of local delegates, who in turn had no power to force compliance with their requests for information.

A "duality of centralization and decentralization" characterized Popular Power.[7] Too often local meetings were no more than the "fulfillment of a liturgy."[8] At their best, they addressed immediate and concrete issues: local Popular Power allowed the citizenry a voice in the conduct of local affairs, a potential arena for self-government. At their worst, local assemblies became rote events that did not empower the citizenry. Moreover, the OPP were inaugurated amid growing economic constraints after a period of relative expansion during the mid-1970s. Popular expectations had closely identified the promise of democratization with improvements in standards of living. Local assemblies, however, did not have the power or the resources to enhance their legitimacy by extending material benefits. Between 1976 and 1984, elections turned over about 50 percent of the delegates. Some 5 to 10 percent of incumbents were recalled during their tenure.[9] High turnover rates could well have been an indication of the vitality of Popular Power because more ordinary Cubans were partaking in public responsibilities. Frequent rotation, however, could also have been a signal that many local delegates had declined renomination because their offices carried much frustration and no power.

Meeting briefly twice a year, the National Assembly had more formal and symbolic purposes. Although it was not a permanent legislature and consequently did not have an actual role in governing Cuba, the assembly regularly heard reports on the provinces and national ministries, and approved annual budgets, economic plans, and a myriad of laws. Often there was debate among the delegates, especially among those with pertinent expertise or historic revolutionary merits. The way the agenda was worded, however, revealed the nature of the debate: discussion and approval of the items at hand. Debate could modify but never reject proposals. The assembly approved most matters unanimously, or nearly so. Yet, the number of delegates participating in discussions increased over time.[10] Invariably, however, once President Castro spoke definitively on an issue, discussion stopped.

Nonetheless, the National Assembly provided a forum for widened elite participation and an avenue for regular disclosure of information to the public. The contrast with the 1960s was quite evident. Sometimes the National Assembly heard singular discussions on the nature of socialism. One of its 1980 sessions witnessed a debate on a proposition to revoke the stipulation that provincial assemblies discuss and approve provincial budgets. In practice, the executive committees discussed and approved annual budgets, and thus the pragmatic solution was to designate them as the appropriate level for presenting the budgets. No one agreed. Vice-President of the Council of State Carlos Rafael Rodríguez voiced the most forceful opposition:

> Practical difficulties should in no way lead us to put principles aside. . . .
> [T]he construction of socialism and communism lays upon us the maxi-
> mum possible participation of all citizens in all aspects of state administra-
> tion. And lays upon us, the maximum participation of all workers in elab-
> orating and implementing the plan. We have to work in that direction and
> whatever we fail to accomplish is a weakness in the functioning of social-
> ism.[11]

Rodríguez likewise warned that because the dictatorship of the prole-tariat could easily degenerate into the "dictatorship of the secretariat," technical imperatives should not compromise democratic principles.

A year earlier, President Castro had addressed the National Assembly in what became popularly known as the *exigencia* (exigency) speech. The transportation minister had earlier suggested that ministers attend local renderings of account of Popular Power to respond to popular complaints. Castro contended that national government officials could not participate in these local meetings without jeopardizing their national responsibilities. These proposed visits would compound, not resolve, problems:

> We are not going to the heart of the matter. . . . We are not dealing with
> our system's—our socialism's—deficiencies. . . . There is a problem of
> *conciencia*. . . . To what extent do we really manifest political, revolution-
> ary, social *conciencia?* We manifest it often . . . incredibly, admirably, ex-
> traordinarily. . . . But, in day-to-day life we are lacking in *conciencia*.[12]

Emphasis on *conciencia* notwithstanding, Castro manifested a concern with procedure and order foreign to the late 1960s. Nonetheless, in his view, the deficiencies of socialism required more conscious cadres at all levels. *Conciencia*, not autonomous institutions and participation, was the essence of good politics.

Popular Power exemplified the politics of Cuban socialism. Local assemblies took public opinion into account more systematically than the politics of mobilization had during the 1960s and thus enhanced popular involvement in the administration of daily life. They did not, however, bestow upon the population the opportunity—let alone the power—to discuss and decide matters of substance. Their mandate was to supervise the state, not to debate investment policies or resource allocation. Involvement—not substantive participation—was the key characteristic of Popular Power at the local level. Moreover, involvement was to be as individuals, not organized groups. Local assemblies were, nonetheless, a significant institutional advance after the debacle of the radical experiment. At the national level, Popular Power allowed for broadened elite participation. Although the Communist party leadership made all fundamental decisions, the National Assembly did discuss issues of substance. Fidel Castro, however, decidedly marked the proceedings: his word was always the last. The politics of "democratic centralism" under which higher institutional levels prevailed over lower ones and the Communist party was the ultimate repository of power characterized the functioning of Popular Power. Moreover, the reality of Cuban socialism added the dimension of charismatic authority.

Institutionalization also entailed economic reorganization. In 1975, the party congress approved the economic management and planning system. JUCEPLAN president, Humberto Pérez—a technocrat without significant credentials in the anti-Batista struggle—spearheaded the implementation of SDPE. The antithesis of the experience of the late 1960s, the new system was an attempt to introduce relative decentralization, profitability criteria, material incentives, and self-financed enterprises. The SDPE instituted financial controls and greater enterprise autonomy, and recognized the role of the law of value—"independently of our will and desires"—in the socialist economy.[13] SDPE implementation, however, moved forward erratically. Making order out of chaos was not easy in the face of continued economic uncertainties, insufficient numbers of trained personnel, and limited political will to assume the full range of consequences of market socialism.

During the late 1970s and early 1980s, the economy began to acquire a new face. With more variety, better quality, and higher prices, the parallel market regularly supplemented rationing for people with higher incomes and for most Cubans on special occasions. Between 1980 and 1986, peasant markets operated and offered a variety of fruits and vegetables the public had not seen in years. By 1979, over 45 percent of workers labored under output norms and quotas, albeit these were unrealistically low and their revisions slow.[14] In 1980–1981, the state implemented wage and price reforms; except in the two Havana provinces, enterprises were permitted to contract labor directly. In 1984, a new law allowed the market to regulate the buying and selling of housing. Daily life in socialist Cuba was assuming less baneful dimensions.

From the outset, the SDPE was in tension with the visions of the 1960s. Although in essence the system repudiated them, the revolutionary experience precluded their overt dismissal. Efficiency and rationality had not inspired the Moncada, the *Granma* landing, the general strike of January 1, the victory at Playa Girón, the Literacy Campaign, nor

Guevara and his comrades in Bolivia. The recognition that material incentives were necessary to motivate a majority of workers was accompanied by the insistence that only moral incentives could counteract individual selfishness. At the 1975 party congress, Fidel Castro warned that the new "mechanisms" were not meant to solve all problems:

> We can[not] do without moral incentives, we would be making a great mistake, because it is absolutely impossible for economic mechanisms and incentives to be as efficient under socialism as they are under capitalism, for the only thing that functions under capitalism is incentive and economic pressure brought to bear with full force, namely, hunger, unemployment, and so on.[15]

The SDPE was not to be a substitute for the party or the state. Politics and ideology were still paramount, and the economy could not be divorced from the legacy of the revolution.

The economic management and planning system undoubtedly operated under special circumstances. A trade-dependent and embargoed economy could not guarantee the flow of resources required by the organization of planning and relative decentralization. Moreover, full implementation of the SDPE carried the danger of broadening inequalities in the population and among regions. Small pockets of unemployment developed as enterprises eliminated underemployment to meet profitability criteria. In other ways, the Cuban experience was similar to that of state socialism. Enterprise autonomy was, for example, resisted by central ministries, and consequently neither self-financing nor improved economic efficiency happened as anticipated. In 1983, an arbitration official in Pinar del Río province succinctly described the SDPE dilemma:

> We evaluate enterprises by the system, but they do not operate according to the system. . . . SDPE mechanisms and resorts are not used. . . . We are still implementing the compulsion mechanism of [material] stimulation. There are no pressures on enterprise managers. They do not bear the consequences of their actions. Nothing happens to them.[16]

JUCEPLAN echoed his complaints.[17] The duality of centralization and decentralization also plagued the SDPE. Although investment decisions were never decentralized, control over the wage fund was. Like other centrally planned economies, the state faced demands for greater investment flexibility from local enterprises and protest from central agencies for excessive salaries.

The Trade Unions as Mass Organizations

The radical experiment had exacted a heavy toll on the trade union movement and the working class. The late 1960s had turned the unions into adjuncts of management and the party. Grass-roots organizations had responded largely to vanguard workers. In 1970, Labor Minister Jorge Risquet had acknowledged that

> theoretically, the administrator represents the interests of the worker-peasant state. . . . Theory is one thing and practice another. . . . The party is so involved with management that in many instances . . . it has become somewhat insensitive to the problems of the masses. . . . If party and administration are one, then there is nowhere the worker can take his problems. . . . The trade union does not exist or it has become a bureau for vanguard workers.[18]

The first step toward the revitalization of trade unions was their reconstruction as mass organizations. *"El sindicato es todos,"* Lázaro Peña told the CTC congress in 1973. Indeed, "the union belongs to all" was potentially more democratic than the appeals to vanguardism and *conciencia* of the late 1960s.

Following the call for democratization in 1970, local elections, conducted by secret ballot rather than acclamation, resulted in a nearly wholesale turnover of trade union incumbents; only 27 percent were reelected. The party subsequently removed from office some leaders "who did not have sufficient merits."[19] By the early 1980s, trade unions functioned in nearly 40,000 enterprises, and elections regularly selected more than 280,000 local leaders.[20] The CTC held congresses in 1973, 1978, and 1984. In 1973 and 1978, rank-and-file representatives accounted for 50 percent and 68 percent of the delegates respectively. In 1973, local leadership turnover figures were not released; in 1978, 54 percent of local trade union leaders were newly elected.[21] The 1984 congress's grass-roots composition was again 68 percent; information on electoral turnover was not available.[22] Nonetheless, CTC General Secretary Roberto Veiga reported that labor leaders with more than 10 years experience had increased from 28 percent to 47 percent between 1978 and 1983.[23] Given the dismemberment of the labor movement during the late 1960s, continuity and experience were notable accomplishments. By the mid-1980s, the CTC faced problems of a different order: responsiveness and accountability to the nearly 50 percent of workers who had been children in 1959 or had been born after the revolution.[24]

Institutionalization meant revitalizing the unions as mass organizations under party leadership. Cuban socialism followed the lead of state socialism: the dictatorship of the proletariat was under the auspices of the vanguard party. In 1973, Raúl Castro underlined the rationale behind party preeminence:

> It is necessary to keep in mind that the working class considered as a whole . . . cannot exercise its own dictatorship. . . . Originating in

bourgeois society, the working class is marked by flaws and vices from the past. The working class is heterogeneous in its consciousness and social behavior. . . . Only through a political party that brings together its conscious minority can the working class . . . construct a socialist society.[25]

Thus, the party guided and directed the unions. Trade unions were a "vehicle for orientation, directives, and goals which the revolution must convey to the working masses" and the "most powerful link" between the party and the people.[26] However, the Communist party was the sole guarantor of socialism.

The 1973 CTC congress elaborated the functions of trade unions as mass organizations marked by "fundamentally cooperative relationships for a superior common objective" with the party, the state, and enterprise management. Each component of socialist society had its own "sphere" and "method" of action. The late 1960s had revealed "grave errors" in the functioning of the unions.[27] Unions were a counterpart to management, and union leaders were charged with defending the "legitimate" interests of workers. "A petit-bourgeois spirit still permeates public administration," Fidel Castro had cautioned in 1970. "An antiworker spirit, a bit of disdain for workers exists among some managers."[28] Although unions were responsible for keeping this antiworker spirit in check, the "superior common objective" of increasing production compelled workers, unions, and management to cooperate. Indeed, interviews with union leaders and rank-and-file workers in 1975 indicated that most defined production as the most important union task. Only 2 in 57 mentioned defense of worker interests. Eight workers referred both to increasing production and defending workers.[29]

Institutionalization similarly did not mean unions had the power to control the economy. Determining plan priorities, the wage fund, and personnel policies were not within union authority. The top echelons of the party and the state decided these substantive matters, and management was entrusted with their implementation. Unions, however, were represented at all levels of the policy-making process. CTC Secretary General Roberto Veiga became a Politburo alternate in 1980 and a full member in 1986. Although trade union leaders accounted for 19 of 148 full Central Committee members in 1980, their numbers fell drastically in 1986 to only 10 of 146 full members.[30] Veiga was also a member of the Council of State and participated in meetings of the Council of Ministers. Trade unions were likewise represented at the provincial and municipal levels of the party and the state. Local union general secretaries were members of enterprise councils. Workers did not elect or recall managers, however. In conjunction with Popular Power, the ministries appointed and dismissed administrators. Although impressionistic evidence suggested workers influenced the dismissal process, influence was a far cry from an established procedure enabling workers to throw out managers who were *hijos de puta,* as Carlos Rafael Rodríguez had expressed it in 1969.[31]

The unequivocal and primary objective of unions was to increase production. Without capitalist exploitation, improvements in living standards depended on economic development. The fundamental activity of the labor movement was "fostering and consolidating the economy." Cuban workers, "keenly aware that we own our national wealth," were more than willing to "sacrifice immediate and particular interests . . . for the benefit of the collective good."[32] Strong unions under vanguard guidance were a requisite for the pursuit of economic development. The economic management and planning system defined a "space" for local enterprises. The SDPE defined the rights and responsibilities of management, workers, and unions.

Workers and the Economy

During the early 1970s, stricter enforcement of labor discipline, establishment of output norms, linkage of wages to performance, and greater availability of goods and services improved labor productivity and rendered material incentives meaningful. In 1971, an antiloafing law contributed to curbing absenteeism. Although the 1973 CTC congress focused largely on economic issues, its documents restored the trade unions under vanguard party leadership. CTC theses and resolutions constituted de facto condemnation of the radical experiment. "From each according to his ability, to each according to his work" clearly departed from the goal of equalizing wages from the bottom up regardless of work performed. The CTC congress also eliminated the so-called historical salaries and full-pay retirement in vanguard enterprises and approved the use of volunteer work only after the normal work load was finished.[33] During the early 1970s, Cuban economic performance improved markedly.

In 1976, the economy again began to slow down. After an all-time high in 1974–1975, sugar prices plummeted and subsequently remained generally low. Worsening hard-currency trade forced a readjustment of development plans. Even though the supply of consumer goods declined, wage raises soon equaled or surpassed productivity increases. New wage policies could not be fully instituted because they were economically "irrational." "Increasing money in circulation without providing an adequate supply of goods and services," Roberto Veiga told the 1978 CTC congress, "would have constituted a step backwards to the situation we faced between 1967 and 1970."[34] "Socialist inflation" and fiscal constraints were, moreover, aggravated by the emergence of unemployment—the *disponibles* (available ones). The state guaranteed laid-off workers 70 percent of their salaries until they found other employment.

Nevertheless, a general wage reform went into effect in 1980. By November 1981, 94 percent of the labor force was benefiting from the reform. The new wage scales widened the ratio from 4.33:1.00 to 5.29:1.00. Using 1977 as the base, salaries increased an average of 15 percent while productivity improved 35 percent. The wage reform stipulated that 15 to 25 percent of salaries be "mobile," that is, dependent on bonuses and other incentive payments.[35] By 1985, slightly more than 1,000 enterprises employing about 1 million workers were creating the year-end bonus funds. In practice, incentive payments constituted only around 10 percent of average basic wages.[36] In 1981, reform of retail prices raised the cost of more than 1,500 products. The average increase on sixty-four sample items listed in the newspaper *Granma* was 60 percent.[37] Ten days after the reform was announced, the internal commerce minister and the head of the State Committee on Prices were dismissed. Students and workers had protested hikes in restaurant prices.[38]

The 1984 CTC congress gathered under what appeared to be relatively auspicious economic circumstances. The economy was registering reasonable growth rates; consumption was likewise experiencing improvement. Moving forward with the SDPE and its "undeniable" accomplishments was the order of the day.[39] As a party meeting on the economy earlier in the year had done, the congress emphasized economic efficiency and narrowing the gap between profitable and unprofitable enterprises. In 1983, unprofitable enterprises had increased their losses, and profitable ones their gains.[40] The delegates discussed the particularly sensitive law granting laid-off workers 70 percent of their salaries; the economy could not sustain these benefits and their future reduction was augured.

Workers and Management

Improving production and defending worker interests were the twin objectives of trade unions. The demise of capitalism allowed for "cooperative relations" between workers and managers. At the 1978 congress, President Castro observed:

> Today a manager does not belong to another class, he is not the workers' enemy; he came forth from the workers' ranks and is friend, relative, neighbor of those who work with him. . . . We have to demand him to be demanding . . . his job is to be demanding and to control.[41]

The withering away of trade unions during the 1960s, however, required that the terms of union-management relations be carefully delineated.

The SDPE, moreover, created the potential for significant tensions. The organization of enterprises on the basis of profitability often resulted in contradictions between workers and managers on work conditions and other matters. The 1978 CTC congress noted:

> Undoubtedly, we need to develop our economy in order to improve our working and living conditions. But differences and even contradictions can arise. In those cases, trade unions are obliged to seek an honest clarification . . . on the basis that the rights of workers be respected. . . . The defense of worker rights, correctly interpreted, strengthens proletarian power.[42]

After the 1973 congress, collective work agreements regulated worker-management relations. Management was bound to enforce safety regulations, maintain worker lounges, and establish vacation timetables. Workers were supposed to be punctual and disciplined, and to care for their work equipment. The implementation of collective agreements was often lax. Vague commitments, weak procedures for determining their breach, and poor publicity of their content among workers were typical difficulties. Occasionally, managers refused to contract the agreements. The 1984 congress was particularly sensitive to violations of safety conditions because job-related accidents were on the rise. The CTC partially attributed the increase to the use of safety funds to meet more pressing production needs. Unions demanded and obtained the nontransferability of funds earmarked for safety equipment.[43] Most accidents, however, were caused by reasons other than the lack of proper equipment, such as ignorance about rules and regulations, worker refusal to use the equipment, and generally indifferent attitudes about enforcing safety regulations on the part of both trade unions and management.[44]

Monthly production and service assemblies were meant to promote worker participation in "the struggle to improve economic efficiency" and to advance their *conciencia* as owners.[45] The assemblies were called upon to check plan fulfillment, analyze production quality, and discuss labor discipline. Trade unions were encouraged to seek worker criticisms. At the same time, union leaders were also expected to educate workers not to "pry into things which are not their concern" and to express "concrete" and "precise" suggestions for solving problems.[46] The 1978 CTC congress noted that monthly assemblies often turned into "meetings in which a mechanical rattling off of figures is presented and where the analysis of fundamental problems is omitted."[47] Two years later, Roberto Veiga sounded a comparable theme:

> There are work places in which workers express their concerns and disagreements in these assemblies . . . and they are not heard by management. . . . [A] climate of malaise and indifference is generated to the considerable detriment of our economic endeavors.[48]

Interviews conducted in 1975 among vanguard workers underscored similar tendencies. Although forty-nine in fifty-seven workers said that management was obliged to consult them about enterprise matters, only thirty-three referred to their input as influential and thirty indicated management had to respond to worker inquiries and suggestions.

Between 1974 and 1978, 85 percent of the labor force participated in assemblies to discuss production plans. Like monthly assemblies, these meetings were plagued with difficulties. Many ministries released only partial information to enterprises. Management often failed to consider worker input. In 1978, Roberto Veiga warned that such practices resulted in the "mere formality of discussing plans with workers and their unions."[49] JUCEPLAN President Humberto Pérez subsequently disclosed salient information on plan discussions. In 1978, 35 percent of all enterprises never held assemblies to discuss the 1979 plan, and only 42 percent revised it by incorporating worker suggestions.[50] The 1980 plan manifested some improvement: only 9 percent failed to discuss the plan and 59 percent included rank-and-file input.[51] Nonetheless, in his main report to the 1986 party congress, Fidel Castro noted that worker participation in the elaboration of plans was just beginning to improve.[52]

The CTC had no formal recourse to obligate management to consider the input of unions and workers. Before an audience of managers, Roberto Veiga noted in 1980: "To a true manager, reliance on the opinions of workers is not just a question of work style . . . of attitude, neither is it a matter of courtesy. It is an indispensable part of the managerial ability of socialist administrators."[53] Meaningful participation was supposed to promote the *conciencia* of workers as owners as well as advance enterprise performance. Good socialist managers needed to acquire *conciencia* of the double function of participation. The 1984 congress was especially critical of the absence of feedback to rank-and-file suggestions that "irritates" workers and "conspires" against the objective of attaining their "active" and "conscious" participation.[54]

The SDPE, however, also created a concurrence of immediate interests between management and unions. Self-financing underscored the interest of workers and managers in enterprise profitability. Individual bonuses and collective funds for social projects, for instance, depended on the generation of "profit." Because bonuses were salary-based, managers received larger stipends than workers. Nonetheless, the SDPE structured a potential collusion of interests among workers, unions, and management. Local control over the distribution of centrally allotted wage funds and the creation of bonuses promoted cooperation between management and labor in enterprise performance. Both had an interest in defeating the resistance of central ministries to enterprise autonomy and in maximizing the resources disbursed locally. Thus, the SDPE reinforced the immediate *conciencia* of workers and managers without also supporting *conciencia* about the national economy and *la patria*.

During the late 1960s, procedures for arbitrating worker-management disputes had weakened. Between 1974 and 1978, work councils handled an average of 80,000 cases a year, 25 percent of which dealt with worker grievances. Yearly numbers had nearly doubled from the early 1970s.[55] The revitalization of unions, the introduction of material incentives, and the establishment of the SDPE caused the increases in labor-management grievances. In 1977, the National Assembly placed the councils under CTC jurisdiction. The 1978 CTC congress pledged to strengthen them as instruments of arbitration and labor justice. Problems of indiscipline and low productivity persisted, however. Wage increases, the *disponibles,* and work stoppages caused by shortages of raw materials were undermining efforts to attain greater economic efficiency. Labor discipline became a central focus of public discussion. In 1979, Fidel Castro told the National Assembly: "Today our labor laws are actually protecting delinquency . . . the lazy, absenteeist worker . . . not the good worker."[56]

In 1980, the Council of Ministers divested work councils of their power to hear labor discipline cases because they were extremely slow in settling disputes and were failing to improve discipline and productivity. Decree No. 32 granted management full authority to enforce labor discipline: managers could now sanction and even dismiss workers. Workers had the right to appeal management actions in municipal courts. The Council of Ministers simultaneously enacted Decree No. 36 to regulate management. Managers, however, were sanctioned by their ministries, not the workers. By 1984, these decrees were deemed highly effective: productivity increases surpassed projected rates.[57] Not surprisingly, Decree No. 32 was initially enforced with vigor and in excess. Not infrequently, management resorted to dismissal as a first sanction against a worker. Also not unexpectedly, Decree No. 36 was unevenly enforced.

The unions, however, implemented some corrective measures. Union inspections and worker appeals attained compensation out of enterprise funds for workers who had been unfairly sanctioned. Although some worker assemblies suggested that indemnification be taken out of manager salaries, their suggestion was unequivocally dismissed.[58] Disciplinary rules and regulations were elaborated to curb management arbitrariness in enforcing labor discipline. Decree No. 36 was more regularly applied, especially against managers who exceeded their authority under Decree No. 32. In 1984, nonetheless, the CTC congress took note of continued worker dissatisfaction with the more lenient application of discipline measures against managers.[59] At no time, however, did the CTC acknowledge that enactment of Decree No. 32 contravened stated intentions of widening worker participation. Granting management full authority over labor discipline did not contribute to empowering workers and fostering in them *conciencia* as owners. On occasion, relations between managers and workers turned less than cooperative as evidenced by the demand that managers pay unjustly disciplined workers out of their salaries. Although it improved labor discipline, the decree was incompatible with the call in 1970 for a collective body to manage enterprises.

During the 1970s and early 1980s, the unions were revitalized under the guidance of the Communist party. The working class bore the burden of legitimating socialism, but workers did not have the power to make national policies. Their charge was to work hard. The Communist party exercised power on their behalf, and Fidel Castro was the premier expositor of their welfare. The correct proletarian *conciencia* was to abide by party directives and charismatic authority. In that sense, Cuban socialism was like the other contemporary socialist experiences: the working class wielded power vicariously.

The Federation of Cuban Women and Gender Equality

Institutionalization brought significant changes to the FMC and Cuban women. Like the CTC, the FMC also celebrated three congresses after 1970: 1974, 1980, 1985. At the 1974 congress, Fidel Castro succinctly stated: "Women's full equality does not yet exist."[60] A year later the party congress formulated an affirmative action policy toward women, pledged to "eliminate all vestiges of the past," and charged the FMC with defending the interests of women.[61]

A crosscut view of the party, mass organizations, and Popular Power in the mid-1970s revealed a modest representation of women leaders. Women constituted 13 percent of party membership; only 6 percent occupied national cadre positions. Six of the 112 full members of the Central Committee were women. No women sat on the Politburo or the Secretariat. In the Communist Youth, women accounted for 10 percent of the national leadership and 29 percent of the membership. Only 7 percent of national trade union leaders were women. With 50 percent of the membership, women held 19 percent of the national leadership of the Committees for the Defense of the Revolution. In Popular Power, women delegates were 8 percent at the local level, 14 percent at the provincial level, nearly 22 percent in the National Assembly. Except for the Communist Youth and the CTC, the policy of affirmative action resulted in more women in national leadership positions than among local cadres (see Table 6.1).

By 1979–1980, the number of women in leadership positions had grown. Available data do not permit exact comparisons with the mid-1970s, but adequate parallels can be drawn. In 1980, women accounted for 19 percent of PCC membership. The party, however, had pledged to match the share of women in the labor force, which was then 32.4 percent. Of 148 full members of the Central Committee, 18 (12.2

Table 6.1. Female Membership and Leadership in the Party, Mass Organizations, and Popular Power Assemblies, Cuba, 1975–1986 (in percentages)

Time Period	PCC	UJC	CTC
1975–1976			
Members	13.2	29.0	24.0
Leaders			
Local	2.9	22.0	24.0
Provincial	6.3	7.0	15.0
National	6.0	10.0	7.0
Central Committee	5.4	—	—
1979–1980			
Members	19.1	41.8	32.4
Leaders			
Local	16.5	n.a.	42.7
Provincial	15.0	n.a.	17.8
National	9.0	14.3	16.1
Central Committee	12.2	26.4	7.7
1984–1986			
Members	21.5	41.0	38.0
Leaders			
Local	23.5	47.6	45.1
Provincial	16.9	28.9	14.7
National	12.8	19.5	17.7
Central Committee	12.3	27.1	2.4

Sources: Primer Congreso del Partido Comunista de Cuba, *Tesis y resoluciones* (Havana: Departamento de Orientación Revolucionaria, 1976), p. 585; *Second Congress of the Communist Party of Cuba: Documents and Speeches* (Havana: Political Publishers, 1981), pp. 66, 74, 78, 415–421; Fidel Castro, *Informe Central: Tercer Congreso del Partido Comunista de Cuba* (Havana: Editora Política, 1986), p. 92; *Cuban Women, 1975–1979* (Havana, 1980), pp. 26, 29; *XV Congreso de la CTC: Memorias* (Havana: Editorial de Ciencias Sociales, 1984),

percent) were women. FMC President Vilma Espín was promoted to alternate status in the Politburo. Forty percent of Communist Youth militants were female. More than 40 percent of local trade union leaders were women. There were slightly more female delegates in the National Assembly (22.6 percent) and slightly fewer in the local assemblies (7.2 percent). By the end of the 1970s, women were increasing their numbers at all levels of these organizations and institutions.

By 1984–1986, women were continuing to make some inroads. At the 1986 party congress, Vilma Espín became a full Politburo member. Two other women were included as alternates. The share of women in full Central Committee membership remained the same as in 1980. Female PCC members increased slightly to 21.5 percent. The proportion of women party cadres rose to 23.5 percent, local Communist Youth leaders to 47.6 percent. The share of women delegates to local (17.1 percent)

Table 6.1. (continued)

	CDR	Popular Power
1975–1976		
Members	50.0	—
Leaders		
Local	7.0	8.0
Provincial	3.0	14.0
National	19.0	21.8
1979–1980		
Members	50.0	—
Leaders		
Local	41.0	7.2
Provincial	31.0	17.4
National	30.0	22.6
1984–1986		
Members	49.4	—
Leaders		
Local	37.5	17.1
Provincial	37.5	21.4
National	31.8	22.4

pp. 268–269; Vilma Espín, "La batalla por el ejercicio pleno de la igualdad de la mujer: acción de los comunistas," *Cuba Socialista* 20 (March–April 1986): 50, 54–55; *Granma*, February 8, 1986, Supplement, and December 29, 1986, p.3; *Granma Weekly Review*, January 4, 1976, p. 12, and November 16, 1986, p. 3; *Bohemia*, November 16, 1976, p. 48, and September 17, 1985, p. 82.

and provincial (21.4 percent) Popular Power assemblies increased but stagnated at the national level. More than 45 percent of local union leaders and about 38 percent of local CDR leaders were women. The relative success of affirmative action reflected party commitment and FMC diligence in pursuing equality. Nonetheless, material and cultural obstacles stood in the way of full equality.[62]

Educationally, as was the case before 1959, Cuban women did not differ significantly from men. By the early 1980s, close to 5 percent of all men and 4 percent of all women had achieved a university degree. Moreover, educational trends pointed to an even greater leveling in the potential pool of women available to assume positions of responsibility. Careers such as economics and engineering had, respectively, 55 percent and 27 percent female enrollment. Women constituted 81 percent of philosophy majors, a politically selective field conducive to cadre positions.[63] In 1986–1987, women accounted for 55.2 percent of total enrollment in higher education.[64] Lack of education was thus not an obstacle preventing women from attaining leadership positions.

There was some evidence, however, that in at least one career access to women was being limited. After 1984, medical school enrollment was subjected to a 52 : 48 ratio of women to men. Without it, women medical students would outnumber men 3 : 2. Civilian medical aid was a crucial component of Cuban foreign policy. Quotas were necessary, Castro argued, for two reasons. Women had greater family and personal responsibilities and found it harder to go abroad for extended periods. Also, the recipient countries had not undergone the changes with respect to the position of women in society that Cuba was experiencing.[65] When in apparent conflict, national goals subordinated particular interests: Cuban foreign policy required male doctors. The state did not allow individual women to make their own decisions about bearing the burden of extended tours abroad or facing sexism in other societies. Medical school quotas contradicted the commitment to equality and opened the possibility that other careers could be limited if national imperatives so dictated. The FMC did not challenge the quota policy. Establishing it had been a matter of national concern that precluded the pursuit of gender equality. Vanguard party politics allowed the FMC to defend the interests of women—like the CTC those of workers—only within its sphere.

In the mid-1970s, the party conducted a survey among 302 men and 333 women in Matanzas. The PCC sought to understand the reasons for the small number of women elected as local Popular Power delegates. When asked why women did not hold leadership positions, nearly 60 percent answered that a woman was responsible for taking care of home, children, and husband. When women were asked about their willingness to serve if elected, 54 percent answered they could not because of family responsibilities. When both men and women were asked why fewer than 10 percent of the candidates had been women, one-third once again pointed to household and child-care obligations. Finally, a question was asked about the personal characteristics expected of a delegate. About 45 percent responded "moral, serious, decent" for women; 20 percent alluded to the same virtues for men.[66] By the mid-1980s, the percentage of women elected as local Popular Power delegates had doubled. Nonetheless, cultural and material factors surely limited women's fuller involvement in Popular Power and other aspects of public life. If the average female worker was also enrolled in an adult education course, was a party and/or trade union activist, and spent over four hours a day on domestic chores, she would have been unlikely to have the time or the disposition to assume additional responsibilities.[67]

After the early 1970s, Cuban women made impressive advances in their access to leadership positions; by the mid-1980s, women held approximately 25 percent of these posts. Between 1968 and 1974, the average had been 6 percent. PCC affirmative action policies and FMC advocacy yielded positive results. Women themselves assumed a more activist stance, as indicated by their willingness to accept positions of responsibility. Holding public office was not tantamount to the exercise of power, however. If they were going to be more than a token presence, women leaders needed to advance the interests of women. And the institutions and organizations in which they were leaders needed to have the power to articulate these interests. The broader issue was the nature of vanguard party politics. During the 1970s and early 1980s, the FMC succeeded in revising employment policies to favor the interests of women workers. The dynamic between the party and the FMC seemed to be more effective for women than that of the party and the CTC was for workers.

Women and Work

After 1970, women significantly expanded their share of the labor force. By 1986, women were 38 percent of the labor force and had attained a notable degree of stability. For every hundred women entering the labor force, fewer than four dropped out.[68] Nonetheless, the 1970s witnessed tensions between national economic prerogatives and the expansion of female employment. The Cuban leadership saw the incorporation of women into the labor force as fundamental to overcoming gender inequalities. The SDPE, however, placed a premium on efficiency and rationality. In the mid-1970s, women accounted for nearly 26 percent of the labor force. The two 1968 resolutions regulating female employment, passed when Cuba faced a rural labor shortage, continued to reserve some jobs for women and proscribe others. The 1974 FMC congress criticized the restrictive resolution and demanded its revision, arguing that prohibition implied discrimination and that women themselves should decide whether or not to perform these jobs.[69] The PCC thesis on full equality supported the FMC position; the SDPE emphasis on efficiency that was creating some unemployment did not.

In 1976, the Labor Ministry passed a new resolution that, contrary to FMC expectations, barred women from nearly three hundred job categories. The ministry allegedly based the selection on the health hazards the jobs presented for women.[70] The resolution, however, had more to do with the problem of unemployment than with the health of women. Banning women from those job categories—in some of which women were then working, and the resolution prescribed their transfer—opened employment opportunities for surplus male workers. Women without jobs did not constitute the same kind of social problem that unemployed men did, or so the resolution seemed to imply. That female employment would stagnate or even decrease was also implicit in the resolution. In 1977, Vilma Espín acknowledged the FMC was seeking its modification.[71] The share of women in the labor force continued to increase. New women entrants, however, tended to have technical, skilled, or professional qualifications.[72] The employment of educated women comple-

mented the national interest. At a time of growing unemployment, however, the state deemed an unqualified policy of equality of employment
unsalutary for the economy.

By the mid-1980s, the original list of "off-limits" job categories had
been whittled down to about twenty-five.[73] The FMC had generally
succeeded in the struggle against job discrimination. Espín, moreover,
reasserted the 1974 FMC position on job prohibitions for women: "The
establishment of prohibitions for women in general is indeed negative,
because they constitute a violation of the principle of equality."[74] Recognizing the controversy, Fidel Castro noted: "If we fall back with respect
to jobs, if we fall back in the economic field, we will start going back on
everything else we have gained."[75] The practical denouement of the
1976 resolution established that women were necessary for the economy, and work was essential for full equality. The FMC successfully
defended the particular interests of women amid pressures to sidetrack
them for the sake of national development. The FMC, however, did not
succeed in having the resolution repealed; its discriminatory intent remained in effect.

In contrast to the medical school quota, the FMC lobbied and won a
de facto victory on the 1976 resolution. In both cases, however, the issue
of gender equality was secondary to the national interest as understood
by the party and state leadership. The quota case affected a smaller
number of women and the very sensitive area of foreign policy that was
a sacrosanct reserve of the top PCC leadership. The FMC did not intervene. More revealing of the potential mass organizations had under the
post-1970 institutionalization was the FMC lobby with respect to the
1976 resolution. The federation argued successfully for the expansion of
female employment—albeit at the skilled, technical, or professional
levels—in the face of increasing unemployment. The specific interests of
women and the SDPE were reconciled. In practice, nonetheless, the
principle of gender equality was not redeemed because even if severely
constrained, the resolution remained in effect.

In the mid-1980s, the question of employment appeared to be pointing in a different direction. The 1976 resolution had sought to alleviate the SDPE-originated unemployment. Declining fertility, however,
pointed to a relative shortage of young workers during the 1990s.[76]
Demographic changes might thus augur a new area of concern for the
incorporation of women into the labor force. Fertility trends, however,
seemed to emphasize the importance of women as childbearers. Would
the state adopt a policy to encourage women to stay home and have
more children? Were that to be the case, how would the FMC react?
Without doubt, new challenges await the FMC. Addressing these challenges on whatever terrain they might arise will surely test the organizational efficacy of the FMC and the commitment to the principle of gender
equality.

At the time, institutionalization highlighted attention to the interests of working women. In coordination with the CTC and Popular Power provincial assemblies, the FMC established commissions to analyze job opportunities for women and supervise hiring practices under the economic management and planning system. The SDPE increased the costs of female employment and the likelihood of discrimination.[77] Women were sometimes considered a hindrance to enterprise "profitability": they were more likely to stay home to care for a sick child or an elderly family member, or to be late because of children and family obligations. Managers were sometimes reluctant to promote qualified women for similar reasons. Although women managers were probably more sensitive to the problems of female workers, they accounted for less than 23 percent of all management posts, and most were in junior positions.[78] In 1981, there were only 246 women enterprise directors (8.7 percent) in a total of 2,815.[79] The 1984 CTC congress rejected a proposal to lower the retirement age for women from 55 to 50. Managers would be all the more hesitant to hire women workers if their retirement were allowed even earlier,[80] but early retirement for women would open up jobs for unemployed men. Evidently, the tendency to increase employment at the expense of women had not receded.

Between 1970 and 1985, the structure of female employment underwent further transformations (see Table 6.2). The proportion of women in agriculture and communications declined slightly and that in commerce significantly. The share of women in industry remained approximately the same. Significantly more women were working in construction and transportation. Nearly half of all working women were employed in the nonproductive services. Women workers were generally better educated than men workers: 46 percent had at least a high school education; only 34 percent of the men did. Women constituted more than 45 percent of the labor force with a high school education or above.[81] Both indices were above their 38 percent share of the labor force. Even so, women tended to earn significantly less than men: 62.6 percent of men were employed in sectors where wages were above the national average of 2,252 pesos; only 38.6 percent of women were so employed.[82]

Women needed an infrastructure of support services in order to work. Although a new *conciencia* in men would also alleviate the overload borne by women, such awareness developed slowly. The 1975 Family Code had stipulated equality between sexes at home and at work, but nearly a decade later Vilma Espín was still emphasizing that men and women were supposed to share child care and household chores: "If we use the term 'help' we are accepting that these are women's responsibilities and such is not the case: we say 'share' because they are a family responsibility."[83] In 1986, Espín also asserted that sharing was a party directive: men who evaded their responsibilities at

Table 6.2. Distribution of Men and Women by Economic Sector,
Cuba 1985

Sector	Salary[a]	Men[b]	Women[c]
Culture	2,689	22,500	17,100
Transportation	2,589	161,800	35,900
Science	2,531	14,400	13,300
Other productive	2,454	7,400	6,400
Construction	2,442	275,700	43,900
Administration	2,404	88,800	73,000
Industry	2,329	660,000	260,000
Other nonproductive	2,257	11,300	10,100
Finances	2,235	5,900	13,400
Education	2,178	127,900	260,300
Agriculture	2,155	249,300	70,300
Communications	2,137	14,400	12,600
Public health	2,124	62,200	139,700
Silviculture	2,120	24,300	5,100
Commerce	2,023	190,800	179,100
Personal services	1,955	67,700	48,700
% Above average salary		62.6	38.6

Sources: Comité Estatal de Estadísticas, *Anuario Estadístico de Cuba,* 1986, pp. 196, 200.
[a] Average salary = 2,252 pesos.
[b] N = 1,983,800.
[c] N = 1,189,500.

home were exploiting women and discriminating against them.[84] Nonetheless, family and household obligations fell disproportionately on
women. One of the most important elements in the support infrastructure was the day-care program. Between 1970 and 1986, day-care enrollment more than doubled, from more than 47,000 to nearly 110,000
children.[85] Demand, however, outstripped existing capacity, and the
economy had limited resources.

The PCC as a Vanguard Party

"Men die, the party is immortal!" exclaimed Fidel Castro in 1974. "The
Party is the soul of the Cuban Revolution," he told the party congress in
1975.[86] The 1980 congress heard him say that the party was "the Revolution's finest expression and guarantee par excellence of its historic
continuity."[87] At the 1986 gathering, he reiterated the centrality of the
party:

> During these years of tense struggle, the party has continued its devel-
> opment as the great force of leadership and coherence in our society. The
> party represents with excellence the authority, the morale, and the princi-

ples of the watchful *conciencia* of the Revolution. . . . [The party] has fulfilled with dignity its responsibility to give always the best example in organization, exigence, determination to improve, discipline, revolutionary
austerity, disposition to sacrifice, and close, permanent bonds with *el
pueblo*.[88]

The process of institutionalization consolidated the idea of vanguardism
that had guided Fidel Castro and the *rebeldes* in the struggle against
Batista, empowered the social revolution against the domestic opposition and the United States during 1959–1961, supported the incipient
order of the early 1960s, and sustained the radical experiment of the late
1960s. Having failed to forge alternate forms of vanguard politics, the
Cuban leadership embraced the process of institutionalization.

Strengthening and broadening the Communist party became central
to the politics of the 1970s. The first step was the activation of party
leadership bodies. During the late 1960s, the Politburo, Secretariat, and
Central Committee had barely functioned. After the early 1970s, they
began to operate regularly and integrated a broadened Cuban leadership. The split between old and new communists started to lose significance. After the debacle of the late 1960s, all tendencies agreed on the
course Cuban socialism was taking: Cuba no longer had the political and
economic resources for advocating a *sui generis* model. Old communists
were reinstated to the Politburo and retained about a 20 percent share of
the Central Committee through the early 1980s. By the 1986 congress,
the historic split was no longer relevant. Old communists were dying,
and the issues that had divided Cuban elites had largely been surpassed.
The politics of socialism was now more important in determining elite
dynamics than the history of revolutionary struggle.

Central Committee composition was indicative of these changing
dynamics. Whereas in 1965 the armed forces and the Interior Ministry
had accounted for 58 percent of CC membership, their share declined
steadily to 17.8 percent between 1975 and 1986. Reduced military
presence—often more formal than substantive because of the transition
many officers made to civilian life—symbolized the emergence of a
broadened governing elite. Representatives of the party apparatus increased from 10 percent to 28.6 percent between 1965 and 1975, declining sharply in 1980 (20.3 percent) and gaining again in 1986 (24.7
percent). State functionaries were about 17 percent of the Central Committee until 1986, when their share increased to 26 percent. The mass
organizations experienced marked fluctuations: about 6–7 percent in
the first two Central Committees, nearly 20 percent in 1980, and down
to 13 percent in 1986. After 1975, individuals working in other sectors—
most of whom were ordinary citizens—hovered around 15 percent (see
Table 6.3). In 1986, the category of member of the *Comandante en Jefe's*
advisory commission was introduced. Under charismatic authority, the
process of institutionalization included innovative bodies in addition to

Table 6.3. Central Committee (Full Membership), Cuba, 1965–1986
(in percentages)

	1965	1975	1980	1986
Total	100	112	148	146
PCC	10.0	28.6	20.3	24.7
State	17.0	17.9	16.9	26.0
Military	58.0	32.1	24.3	17.8
Mass Organizations	7.0	6.3	18.9	13.0
Other	8.0	15.1	19.6	13.7
Advisory Commission	—	—	—	4.8

Sources: Jorge I. Domínguez, *Cuba: Order and Revolution* (Cambridge: Harvard University Press, 1978), p. 312, and "The New Demand for Orderliness," in Jorge I. Domínguez, ed., *Cuba—Internal and International Affairs* (Beverly Hills: Sage, 1982), p. 24; 1986 figures computed from *Granma,* Supplement, February 8, 1986.

the orthodox structures of vanguard parties. The more formal bodies did not fully meet the needs of Fidel Castro for governing Cuba. Real elite turnover did not occur until 1986, when approximately 50 percent of the Central Committee was newly elected.[89] Before then, the party had expanded the size of the CC to accommodate new members.

Although the weight of the Central Committee in actual policy-making was difficult to determine, the formal appearance of elite politics in Cuba had changed significantly from the 1960s. If only symbolically, varying CC composition was a recognition of the increasing complexity that socialism was forging in Cuban society. That—even at the height of Cuban internationalism—the share of the military continued to decline was illustrative of the weight civilian and domestic imperatives had in the conduct of daily affairs. In 1980, following the Mariel exodus and the Polish Solidarity movement, Cuban leaders saw some reason to be concerned with their relationship to the "masses" and, consequently, the presence of the mass organizations and ordinary citizens in the Central Committee grew. In 1986, however, when elite turnover happened, the beneficiaries were the party and state apparatuses. Thus, the politics of socialism accorded particular importance to PCC cadres and high-level bureaucrats over other sectors. Nonetheless, Cuban socialism never fully functioned like state socialism: the *Comandante en Jefe*'s advisory commission was the foremost indication of its distinguishing characteristics.

The relationship between vanguard parties and the populations they claimed to represent was always indirect. Popular elections did not mediate the selection of national leaders. Rather, the presence of the vanguard throughout society and the profile of its members supposedly constituted the guarantee of responsiveness to popular interests, especially those of the working class. Initially, the PCC experienced slow growth and then stagnation. In 1969, the party began a period of rapid expansion, and a year later membership totaled about 100,000 (1 percent of the population). During the 1970s, the party underwent extraordinary growth: membership more than quadrupled. In 1980, militants numbered 434,943 (4.5 of the population). Like the Central Committee, the number of members remained fairly stable between 1980 and 1985. At 523,639, membership grew about 20 percent (5.2 percent of the population). Rank-and-file turnover, however, was also significant. In 1985, 39 percent had been in the party for five years or less.[90] After the late 1960s, the educational levels of party members notably improved. In 1975, a majority of party cadres had achieved junior high school, but more than 60 percent of the membership had only a primary education.[91] By 1985, nearly 75 percent of the membership had at a minimum finished the ninth grade and most party cadres had some university education.[92] Contrary to the 1960s, party cadres and members now had the educational qualifications to govern.

After the sectarianism crisis in 1962, the party had adopted the method of selecting members from among vanguard workers. Although the vanguard-worker method remained a path to party militancy after 1975, the Communist Youth increasingly became the standard avenue for PCC membership. In 1985, nearly 60 percent of the party members entered through the Communist Youth. Final approval for party and youth membership nonetheless required "consultation with the masses."[93] Moreover, PCC policy emphasized growth among production workers. Progress, however, was erratic. Production workers represented 30.2 percent (1975), 39.8 percent (1980), and 37.3 percent (1985) of PCC militants. Service workers had a similarly variable record. In contrast, professional/technical personnel and administrative workers experienced steady increases in their share of PCC composition (see Table 6.4). The relative decline of political and administrative cadres from 42.1 percent in 1975 to 23.7 percent in 1986 was notable.

Comparisons between PCC social composition and presence among the different groups in the state civilian labor force further underscored the problem of remaining the vanguard of the working class while other sectors were better represented in PCC ranks (see Table 6.5). Between 1975 and 1985, the share of production workers in the labor force declined slightly. The number of party members among them nearly tripled but PCC workers were still less than 13 percent of the labor force. Service workers' share of the labor force first declined and then increased; their PCC proportion followed a pattern similar to that of production workers. Professional/technical personnel and administrative workers experienced increases in their share of the labor force, and more of them entered the party. Although administrative workers moderately in-

Table 6.4. Social Composition of PCC Membership

	1975	1980	1985
Total membership	211,642	434,943	523,639
Production workers[a] (%)	30.2	39.8	37.3
Service workers[a] (%)	5.7	7.5	5.9
Professional and technical (%)	9.2	15.0	16.5
Administrative cadres[b] (%)	33.4	23.6	20.7
Political cadres[b] (%)	8.7	4.3	3.0
Administrative workers[b] (%)	4.1	4.3	7.2
Peasants[b] (%)	1.8	1.2	2.0
Others[b] (%)	6.9	4.3	7.4

Sources: Primer Congreso del Partido Comunista de Cuba. *Tesis y resoluciones* (Havana: Departamento do Orientación Revolucionaria, 1976), p. 23; Isidro Gómez, "El Partido Comunista de Cuba" (Paper presented at the seminar of the Institute for Cuban Studies, Washington, DC August 16–18, 1979, p. 28; Fidel Castro, *Main Report: Second Congress of the Communist Party of Cuba* (New York: Center for Cuban Studies, 1981), p. 27; Massimo Cavallini, "La revolución es una obra de arte que debe perfeccionarse," *Pensamiento Propio* (May–June 1986): p. 42.

[a] In 1975 and 1980, production and service workers were reported jointly—35.9 percent and 47.3 percent respectively. I estimated the breakdown based on the 1985 figures which were given separately.

[b] I estimated 1980 percentages of these categories based on the Gómez 1978 figures.

Table 6.5. Occupational Distribution of State Civilian Labor Force, Cuba, 1975, 1980, 1985

	1975	1980	1985
Workers (%)	1,343,300	1,354,300	1,604,400
	(56.7)	(52.1)	(50.6)
Services (%)	378,200	348,000	431,400
	(16.0)	(13.4)	(13.6)
Professional and technical (%)	314,500	484,500	635,100
	(13.3)	(18.6)	(20.0)
Administrative (%)	125,700	180,300	248,500
	(5.3)	(6.9)	(7.8)
Cadres (%)	207,600	232,800	253,900
	(8.8)	(9.0)	(8.0)
Total	2,369,300	2,599,900	3,173,300

Sources: Comité Estatal de Estadísticas, *Anuario Estadístico de Cuba 1979,* p. 58, and *Anuario Estadístico de Cuba 1986,* p. 205.

creased their share of the labor force, their presence in the PCC advanced fastest and steadiest to about 15 percent. Between 1975 and 1985, cadres accounted for 8 to 9 percent of the labor force. At each party congress, about half of all cadres were PCC militants; they had the highest proportion of party members in relation to their group totals (see Table 6.6). The vanguard party of the working class was thus becoming more representative of other sectors.

By the mid-1980s, the PCC had acquired the basic profile of the old Communist parties in the Soviet Union and Eastern Europe. Party membership had expanded significantly and, thus, more ordinary citizens were involved in the daily conduct of PCC affairs in their workplaces and neighborhoods. During the 1970s and early 1980s, moreover, the Cuban economy experienced modest growth, and living standards improved noticeably from their trough of the late 1960s. The accomplishments of socialism were beginning to legitimate the rule of the Communist party. The legacy of revolution, however, still weighed significantly, and central to that legacy was the authority of Fidel Castro.

Table 6.6. PCC Members as Percentage of Total in Occupational Categories, 1975, 1980, 1985

	1975	1980	1985
Workers	5.0	12.8	12.2
Services	3.2	9.4	7.2
Professional and technical	6.2	13.5	13.6
Administrative workers	6.9	10.4	15.2
Cadres	42.9	52.1	48.9

Sources: Primer Congreso del Partido Comunista de Cuba, *Tesis y resoluciones* (Havana: Departamento de Orientación Revolucionaria, 1976), p. 585; *Second Congress of the Communist Party of Cuba: Documents and Speeches* (Havana: Political Publishers, 1981), pp. 66, 74, 78, 415–421; Fidel Castro, *Informe Central: Tercer Congreso del Partido Comunista de Cuba* (Havana: Editora Política, 1986), p. 92; *Cuban Women, 1975–1979* (Havana, 1980), pp. 26, 29; *XV Congreso del la CTC: Memorias* (Havana: Editorial de Ciencias Sociales, 1984), pp. 268–269; Vilma Espín, "La batalla por el ejercicio pleno de la igualdad de la mujer: acción de los comunistas," *Cuba Socialista* 20 (March–April 1986): 50, 54–55; *Granma*, February 8, 1986, Supplement, and December 29, 1986, p. 3; *Granma Weekly Review*, January 4, 1976, p. 12, and November 16, 1986, p. 3; *Bohemia*, November 16, 1976, p. 48, and September 17, 1985, p. 82; Comité Estatal de Estadísticas, *Anuario Estadístico de Cuba, 1986*, pp. 196, 200.

Crossroads at Three Party Congresses

The first congress of the Communist party adopted a political and economic program very different from the radical experiment. With institutionalization, the PCC had acquired all the trappings of a vanguard party. Popular Power assemblies were about to be constituted, and a referendum would shortly approve the new constitution. The economic management and planning system of relative decentralization and material incentives most clearly embodied the retreat from the late 1960s. The 1975 congress also underscored a broadened unity among elites: old and new communists came together in what turned out to be the demise of their historic split. Cuban leaders likewise felt satisfied about their relationship to *el pueblo cubano*. Mass organizations were attaining their "proper" level of functioning in a vanguard-led political system. Popular Power assemblies would give citizens the opportunity to voice their immediate concerns. Moreover, the economy had at last registered respectable growth. The year 1975 was a good one.

The Cuban leadership could also look outward with satisfaction. Guerrilla movements in Latin America had faltered, and in 1970 the Chilean electorate—not force of arms—had finally broken Cuban isolation. In 1971, Salvador Allende had warmly welcomed Fidel Castro on an extended visit to Chile—his first to Latin America in more than a decade. Between 1970 and 1975, eight countries in Latin America and the Caribbean established diplomatic relations with Cuba. While shunning the Organization of American States, Cuba became active in other regional organizations. The impact of the U.S. embargo appeared to be lessening. The Cuban government also became more active in the Non-Aligned Movement, and Fidel Castro promoted the idea of confluent interests between the Soviet Union and the Third World. In 1975, with Soviet logistical support, Cuban and Angolan government troops scored an extraordinary victory against the forces of rival Angolan groups and South Africa. In 1977, with even greater Soviet support, the Cuban armed forces came to the aid of Ethiopia when Somalia invaded the Ogaden Desert. Cuban prestige in the Third World was never stronger.

After 1968, relations with the Soviet Union and the other socialist countries improved noticeably, and a year later Soviet economic aid began to increase. CMEA membership secured long-term commitment to Cuban development. The Cuban leadership accepted the new dependence as the price for maintaining socialism ninety miles from the United States. The Soviet Union, moreover, renewed its assurances on Cuban defense, increased the supply of armaments, and upgraded the quality of military aid. In 1972, all the capitals in Eastern Europe welcomed Fidel Castro. Between 1972 and 1975, Castro reciprocated by hosting his fellow communist leaders. In early 1974, Leonid Brezhnev visited Cuba and addressed a large congregation of citizens in the Plaza of the Revolution. The Soviet leader emphasized the soundness of the new course in Cuban domestic and foreign policies. In symbolic recognition of the broadened Cuban leadership, Brezhnev concluded by jointly raising the arms of Fidel and Raúl Castro.

The Cuban-Soviet rapprochement was complete. Although under terms radically different from those anticipated during the late 1960s, Cuban economic and security needs were apparently satisfied. The Soviet Union saw in Cuba a valuable link to Latin America and the Third World. Growing relations between Cuba and Latin America, the defeat of the United States in Vietnam, and the demise of the Nixon presidency also helped to convince the Soviet Union of Cuban commitment to the emerging détente between the two superpowers. In 1973, Cuba and the United States signed an antihijacking agreement. Although prospects for improved relations dimmed with the Angolan expedition, the United States did not pose the same threat to a more consolidated Cuban government that it had during the early years of the revolution. Indeed, the 1975 party congress gathered in good times. Whatever qualms some Cuban leaders—especially perhaps Fidel Castro—might have had about the new course, the promise of success at home and abroad probably assuaged them.

Sending troops to Angola underscored a series of characteristics of Cuban politics after 1959. Named Operation Carlota after a rebellious slave in the nineteenth century, the Angolan expedition was a Cuban initiative possible only because of Soviet support. To a significant degree, Cuban foreign policy attained an extraordinary triumph because the Vietnam War and Watergate had left the United States in a relatively weakened position. Ultimately, Fidel Castro and his closest associates made the decision to send troops to Angola. Elite opposition was highly unlikely; popular support was assumed and probably materialized after the fact. After so many setbacks in Latin America, victory in Angola replenished national pride. No other leader except an audacious visionary like Fidel Castro would have been likely to take advantage of the opportunity that the special circumstances of the mid-1970s afforded. Because of him, Cuba enjoyed the prestige and the security that—at least for a time—resulted from an activist foreign policy.

Involvement in Angola also highlighted the primacy of political over economic considerations in Cuban politics since 1959. The economic costs of Operation Carlota and what turned out to be a fifteen-year stay in Angola were secondary to the goals of internationalism and greater maneuverability for Cuba in world affairs. Similarly, these higher national objectives obscured the multiple tolls support for the Angolan government would take on ordinary citizens. By all accounts, nonetheless, most Cubans who served in Angola did so voluntarily, with great distinction, and notable valor. The Angolan chapter of Cuban foreign policy focused once again on the ability of the leadership to mobilize the population to answer extraordinary challenges. More notable was thus the failure of Fidel Castro and the PCC to engage the citizenry to meet the exigencies of daily life and work.

When the second congress met in 1980, the outlook was considerably less auspicious. After 1976, the economy fell into recession, and expectations for rapid improvements in living standards were disappointing. Implementing the SDPE was more complex than anticipated. The trade-dependent economy—particularly when sugar prices fell—limited the imports needed to support more autonomous enterprises. The political will to assume the consequences of the SDPE was also weak. The problem of the *disponibles* and the relative decentralization of authority ran counter to the Cuban experience. Unemployment before 1959 and the social revolution had rendered full employment and centralized authority two of the principal mainstays of Cuban socialism. The answer to waning efficiency and growing corruption was, thus, to exercise greater *conciencia*. In troubled times, Cuban leaders, especially Fidel Castro, resorted to the recourse that their own experience had nurtured. The material incentives and market mechanisms of the SDPE were not an easy appeal within that experience.

The year 1980 tested the Cuban government like no other since 1970. In April, 10,000 Cubans flocked to the Peruvian Embassy. Between April and September, 125,000 Cubans left Cuba via the Mariel boatlift. The unrealized prospects of the 1970s and the visit of more than 100,000 Cuban-Americans in 1979 had fueled a tense situation. The government labeled those wanting to leave "scum" who renounced the ideals of *la patria* for the lures of consumerism. The PCC organized *mítines de repudio*—meetings to repudiate the "scum"—in front of the homes of those intending to leave. Two decades after the revolution, there was still no room for dissent: *Con Cuba o contra Cuba* continued to define Cuban politics. Ninety miles away from the United States and the prosperous Cuban-American communities, the Cuban government surely had to contend with unreasonable comparisons and inordinate expectations. Still, the challenge for Cuban leaders lay in satisfying basic needs—especially in the supply, diversity, and quality of food and other consumer nondurables—more efficiently, and they had barely met it. Happening at the same time as the Solidarity movement in Poland, the Mariel exodus impressed upon the Cuban leadership the need to reinforce its links with *el pueblo cubano*.

Internationally, Cuba also faced mixed prospects. Relations with the Soviet Union continued on terms largely beneficial to the Cuban government. The election of Jimmy Carter in 1976 created new opportunities for dialogue with the United States. In 1977, interests sections opened in Havana and Washington, and rapprochement proceeded slowly but significantly. Regular communications on varied topics, the end of the U.S.

ban on travel to Cuba, and the release of 3,000 political prisoners were among the most notable accomplishments. In 1979, when Cuba hosted the summit of the Non-Aligned Movement, Fidel Castro became its president. By then, Cuba had an impressive network of military advisors and civilian missions in numerous Third World countries. In September, President Castro addressed the United Nations on behalf of the non-aligned nations. Cuba, however, was not elected to the UN Security Council, as would have been expected because of its presidency of the Non-Aligned Movement. In December, the Soviet Union invaded Afghanistan, and Cuba joined the minority in the United Nations against Soviet censure. In Latin America, relations with Venezuela, Colombia, and Peru deteriorated while they grew with the revolutionary governments of Grenada and Nicaragua. In Maurice Bishop and the Sandinistas, the Cuban government at last had truly kindred allies in the Western Hemisphere.

With the 1980 inauguration of Ronald Reagan, a hostile U.S. administration again confronted Cuba. Although the interests sections remained open, much of the progress made during the Carter presidency was lost. The Reagan administration reinstated the travel ban thereby prohibiting U.S. citizens' travel to Cuba at a time when tourism was emerging as an important source of hard currency. The threat of military aggression also loomed large. The U.S. obsession with Central America included frequent references about going to the "source" of outside intervention, and the Cuban government turned to the organization of popular militias. Institutionalization notwithstanding, Cuban politics could not remain "normal" for long: the Unitd States contributed to maintaining the politics of mobilization and charismatic authority. In 1984, the two countries nonetheless reached an important immigration agreement. Cuba, however, suspended it a few months later when the Reagan administration began the transmissions of Radio Martí.

On balance, relations with Latin America remained good. In 1982, the Cuban government supported the Argentine generals in the war over the Malvinas Islands: Latin American unity took precedence over ideological differences. Cuba also launched a campaign against paying the Latin American debt, which did not gain much official support but garnered the endorsement of numerous opposition groups, intellectuals, social movements, and religious base communities throughout the continent. In 1983, the overthrow of Maurice Bishop and the subsequent U.S. invasion of Grenada were setbacks for Cuban foreign policy. An ally was gone, and Cuban military personnel stationed in Grenada had not fought the U.S. invaders very forcefully. The response of Cuban construction workers gave the Cuban leadership cause to reconsider the established doctrine of national defense. A professional military alone would never suffice: national defense ultimately depended on *el pueblo cubano*, and thus the militias were its first rampart.

The early 1980s, saw greater economic liberalization in Cuba since 1968, when the revolutionary offensive had eliminated the last vestiges of private enterprise. Peasant markets, arts and crafts fairs, self-employment, and a housing market gave the population opportunities to earn and spend more. Outright corruption and what the Cuban leadership deemed to be corrupt practices undoubtedly spread. Many functionaries exploited the perquisites of office for personal gain. Ordinary citizens took full advantage of the market to make money, and goods and services were often sold at exorbitant profit. Many workers and managers used factory inventories for private gain. As early as 1982, Fidel Castro had begun to revive the idea of a communist *conciencia* in the construction of socialism.

The Cuban economy, moreover, was once again facing deteriorating international conditions. Foreign exchange earnings were declining, trade deficits were growing, hard-currency debt was stringently renegotiated, and the prospects for the same levels of trade, credit, and aid from the socialist countries were dim. When the Central Group was created to supervise the economy in 1984 and Humberto Pérez was removed from JUCEPLAN in 1985, the defeat of the economic reformers was imminent. The SDPE and its prescriptions of greater decentralization and material incentives were running counter to the legacy of revolution. *Conciencia* and charismatic authority did not. When the Communist party congress met in 1986, winds of change were decidedly stirring. Although Fidel Castro told the February session that the SDPE encouraged capitalist solutions to the problems of socialism, he did not offer an alternative.[94] The concluding session in December, however, sanctioned a new program of moral renewal and economic restructuring: a process of rectification that Castro had launched in April. That the rectification was his initiative and not the Communist party's was testimony to the relative weakness of the process of institutionalization. In the Soviet Union, the reform program of Mikhail Gorbachev was then in its incipient stages.

CHAPTER 7

Revolution, Rectification, and Contemporary Socialism

A communist spirit and *conciencia,* a revolutionary will and vocation were, are, and always will be a thousand times more powerful than money!

Fidel Castro
December 2, 1986

In these times of confusion in which our Revolution—so feared by reactionaries all over the world, so feared by the empire—stands like a beacon of light. . . . On this January 1 . . . we are aware of the enormous responsibility that our Revolution has with all the peoples of the world, with all the workers of the world, and especially with the peoples of the Third World. We will always meet our responsibility. That is why today we say with more force than ever: Socialism or death! Marxism-Leninism or death! which is now the meaning of what we have repeated so many times over the years: *¡Patria o Muerte! ¡Venceremos!*

Fidel Castro
January 1, 1989

Although the rectification process of Fidel Castro and the reform program of Mikhail Gorbachev coincided in time, their differences were profound. Whereas the Gorbachev reforms aimed—and ultimately failed—to maintain central planning while gradually implementing far-reaching market reforms and to institute an unprecedented political pluralism in vanguard party politics, the rectification evoked the radical experiment of the 1960s. Although Cuba was no longer in revolution, and Cuban society manifested many of the same problems afflicting state socialism elsewhere, Cuban conditions—particularly the leadership of Fidel Castro and continued U.S. hostility—precluded a political opening and economic decentralization.

After 1986, the Cuban government confronted mounting and multiple crises. With the downfall of Eastern European communism and the

153

disintegration of the Soviet Union, it lost the economic lifeline that had permitted survival against the U.S. embargo. In the aftermath of the cold war, the Cuban government was without allies and facing an ever-more-determined antagonist in the United States. Nonetheless, domestic problems prominently underscored the post-1986 crises. Daily life, particularly after 1990, when Soviet trade, credits, and aid dwindled, was becoming extraordinarily burdensome. Moreover, the dynamic of Fidel-*patria*-revolution was wearing down as *el pueblo cubano* was decidedly more concerned about the mundane than about the responsibilities of Cuban socialism before history and the world.

The Process of Rectification

During the mid-1980s, the crisis of socialism fully ripened in Cuba. Socialism had brought the Cuban people national dignity and social justice, but it had been considerably less successful in promoting economic growth and formal democracy. Nonetheless, the Cuban government was not like the governments of the Soviet Union and Eastern Europe. It had come to power by virtue of a genuine social revolution that was still more than a distant occurrence. U.S. hostility, moreover, had sustained the vitality of radical nationalism: social justice and national sovereignty continued to sway significant sectors of the population in support of the government. In contrast to Eastern Europe, the forces of nationalism reinforced socialism. In addition, Fidel Castro remained at the center of Cuban politics. The process of institutionalization had not succeeded in routinizing the dynamic of Fidel-*patria*-revolution.

Competing models of socialism had guided the organization of central planning and one-party politics. More predominant during the 1960s, the first built upon the experience of the social revolution—charismatic authority, popular mobilizations, moral incentives, national independence—to strive for a *sui generis* socialism. Following the debacle of 1970, the second model turned to the practice of state socialism in the Soviet Union and Eastern Europe. Institutionalization, relatively less centralized planning, material incentives, and dependence on the socialist countries characterized the 1970s and early 1980s. Because of U.S. besiegement, the relative nearness of the social revolution, and the continued importance of charismatic authority, Cuban socialism never quite fit the orthodox mold, however.

By the time the Communist party congress convened in February 1986, the second model was largely exhausted. Change—though not in the direction of *perestroika* and *glasnost*—was, indeed, in the offing. Under the leadership of Fidel Castro, the Cuban government did not have the option to embark upon challenges to vanguardism and central planning. Between April and the December session of the PCC congress,

Fidel Castro launched the process of rectification.[1] Major policy initiatives were still his jurisdiction: the PCC lacked the institutional wherewithal to change the course of Cuban politics. To a significant degree, the party still depended on Fidel Castro and thus could not challenge his authority nor the logic of revolution that he sought to perpetuate. Neither was compatible with a political opening and market reforms.

Domestic and international factors, moreover, reinforced the turn toward the rectification. The implementation of the SDPE had reached a crossroads: broader and more meaningful application of market mechanisms, or retrenchment. Growing inequalities, corruption, and a corrosion of *conciencia* at a time when Cuba faced the administration of Ronald Reagan and the original revolutionary leadership was still at the helm weighed decidedly in favor of retrenchment. The rectification curbed the peasant markets and other modest market reforms and reaffirmed the centrality of the Communist party. The new directions disallowed the institutionalization of the 1970s and early 1980s as "bad copies" of the Soviet Union and Eastern Europe. Moreover, hard-currency debt, trade deficits, and a weakening relationship with the socialist countries signaled the urgency of finding ways to improve economic performance. Faced with the crisis of socialism, the Cuban government turned to the tenets of revolution.

The social dynamics of revolution were, however, largely in the past. That Fidel Castro and the historic leadership still retained the reins of power allowed the use of revolutionary rhetoric to attempt to mobilize the citizenry. That the 1960s were sufficiently near in time and U.S. hostility continued relentless infused the idea of the revolution with passion and commitment for many citizens. Still, Cuba was no longer in revolution, and the reality of socialism preeminently determined the consciousness of daily living. Moreover, a rapidly changing world challenged the restoration of the old visions. Without the Soviet Union, the future of socialism in Cuba was uncertain. The crucial question was whether the resurrection of revolutionary visions would succeed in engaging the citizenry of Cuba—an overwhelming majority of whom were born or grew up after 1959—in renovating Cuban socialism. And the rectification, indeed, aimed at nothing less.

The Economics of Rectification

That crisis beset the Cuban economy during the mid-1980s was indisputable. The Cuba-USSR relationship had reached a threshold. Mounting hard-currency debt was rendering Cuba more vulnerable to the world economy. Because of the U.S. embargo, the Cuban government did not benefit from the devaluation of the dollar and was in fact hurt by it because the costs of its hard-currency imports went up. Sugar prices in the world market fell to the levels of the 1930s, and consequently hard-currency earnings declined 50 to 60 percent. Moreover, Western creditors exacted considerably harsher repayment terms from Cuba than from other Third World countries. Without doubt, austerity policies were needed to promote domestic savings and improve efficiency.

Quite clearly, rectification was not the only option open to the Cuban government to confront the critical juncture of the mid-1980s. A reform program of widening market relations and better enforcement of economic controls might have been formulated to mitigate the problems of the SDPE. Rather than closing the peasant markets, for example, the government might have levied higher taxes on profits and more effectively regulated the distribution of produce to urban areas. However, Cuban leaders, especially Fidel Castro, considered that such measures and the material incentives they implied compromised the ideal of socialism. The logic of revolution emphasized reliance on *conciencia* and mass mobilizations as levers to achieve savings and improve productivity. Cuban leaders assessed the consequences of the SDPE to have been inimical to the popular resolve and national unity that had made survival against the United States possible. At a time of growing austerity, they could not fathom calling for the enrichment of some Cubans at the expense of others.

In July 1986, the government issued a report on the problems that the rectification aimed to correct. It documented widespread violations of regulations and lack of control over the economy. Work norms were outdated, salaries incommensurate to output. Marginal production—originally meant to maximize use of residual materials—had superseded primary production in many enterprises. Many workers received a full day's pay for half a day's work for the state and spent the afternoons pursuing their private gain at other jobs. Too often managers contracted skilled labor at higher than prescribed wages without subsequently enforcing labor discipline to increase productivity. The investment process was chaotic and wasteful. Frequently, many enterprises did not enforce their budgets; sometimes, they never developed one. State inspections were generally ineffective. Interenterprise contracts were underutilized and loosely regulated. Management regularly inflated prices to meet output in value without regard to the quality of production. The planning process demanded inordinate amounts of paperwork and little attention to worker input.

The government report also highlighted the political consequences of these problems. Economism was pervasive. The pursuit of individual gain—excessive and often illegal—was the overriding concern of many people. Economic mechanisms and material incentives were displacing *conciencia*. Volunteer work had all but disappeared. Nepotism and *sociolismo* flourished. When transferred to new positions, cadres frequently hired their friends and relatives. Workers who denounced corruption often found themselves sidelined. When enforced, sanctions were more common against people in lower levels of authority. Managemment usually failed to involve workers in solving problems. State functionaries manifested disdain and disregard for public opinion and too often used state resources for private purposes. Too many enterprises showed discrepancies in cash transactions.[2]

Most of these problems were not new. Investments had never been efficient and orderly. Planning had always tended to be bureaucratic, worker participation quite modest, and labor discipline problematic. The radical experiment had especially disregarded the conduct of day-to-day affairs and the input of ordinary workers. How much respect and regard the Cuban government had had for public opinion was a matter of some contention. *Sociolismo* and the private use of state resources had likewise been commonplace. Other problems were more directly the consequence of the SDPE. Local control over the wage funds had created the possibility of excessive salaries. Greater availability of goods and services had enhanced the value of money and the likelihood that people would seek ways to earn more. The state never fully defined SDPE mechanisms for budgets, contracts, financing, and other measures of enterprise performance. Between 1979 and 1985, JUCEPLAN had emphasized most of these problems in four reports on the SDPE. Indeed, the rectification assumed ongoing criticisms of the SDPE while rejecting the prescription of further decentralization.

After 1984, the party began to curtail the powers of JUCEPLAN. The Central Group took over many of its functions, and Humberto Pérez was demoted. In late 1984, Castro had first emphasized the need to engage the people in the struggle for the economy. *Conciencia* had made possible renewed mobilizations for national defense in response to a threatening Reagan administration. The revitalized militias gave testimony to the popular involvement that had marked the best years of the revolution. *Conciencia*, the Cuban leadership once again contended, was also an economic recourse.[3]

Not until 1986, however, did the 1984 emphasis become policy. The rectification disregarded further market reforms and restricted the operation of the SDPE. Curbing salaries, revising norms, and other measures aimed to improve enterprise profitability and save up to 500 million pesos a year.[4] The microbrigades absorbed surplus labor and constructed day-care centers and housing worth 300 million pesos in 1988.[5] In December 1986, the government announced an austerity program under the banner of equality. While the prices of public transportation, utilities, and parallel market goods increased, the program cut back more heavily on the perquisites of state functionaries.[6] Shortly thereafter, workers whose wages ranked them in the lowest 10 percent of the labor force received salary increases of 10 to 18 percent.[7]

Until 1989, the rectification process sought to improve the efficiency of the Cuban economy under international conditions that had not yet radically changed. The new directions aimed to mobilize *conciencia* in pursuit of economic objectives and regain the national *imprimatur* for socialism in Cuba. While praising the radical experiment of the 1960s over the institutionalization of the 1970s and 1980s, the Cuban leadership underscored the importance of avoiding the "idealistic" mistakes of the past. Nonetheless, the policies of promoting efficiency by curbing market mechanisms and appealing to *conciencia* raised serious questions. These policies had not worked when the effervescence of revolution was a more recent experience, and there was little reason to expect that they would work when the reality of socialism had already marked popular awareness. Moreover, after 1989, the Cuban government suffered a drastic reduction of the network of trade, credits, and aid with the Soviet Union and Eastern Europe. In addition, the collapse of state socialism discredited the model of one-party politics and centrally planned economies.

In the early 1990s, Cuban leaders were increasingly facing a situation not unlike the one they would have faced in the early 1960s had the Soviet Union not been willing to become their mentor. How to survive when the international infrastructure of the past three decades was rapidly disintegrating and the United States now more than ever persisted with the embargo in a final effort to force them to capitulate? More starkly than in 1986 when the rectification was launched, the choices after 1989 were between accepting the need for widespread market reforms or insisting on policies aimed at forging a rather elusive "efficient socialist organization."[8] Generally speaking, the latter was the option.

In August 1990, when the government declared the special period in peacetime, it signaled a commitment to socialism against all odds.[9] The special period was an attempt to reinsert the Cuban economy into the world economy without relinquishing socialism and compromising national sovereignty to the United States. If successful, such a transition would, indeed, be as extraordinary as that which Cuba had achieved during the early 1960s, when the erstwhile socialist countries had supplanted the United States. To succeed, the socialist economy had to sustain levels of efficiency that, even under more amenable international circumstances, it had never attained. Labor productivity and the incentives to work under socialism were crucial questions.

When launched in 1986, the rectification confronted the long-standing dilemma of the working class under socialism. The experience of state socialism indicated that governing in the name of the proletariat did not truly empower workers. Party and state interests were clearly not the same as those of the working class. Moreover, workers customarily manifested an awareness of their interests much more bound by the here-and-now than the historically minded (in the best of cases) leaders of vanguard parties. Quotidian concerns were much more likely to motivate the working class than the inevitable march of history. The distinctiveness of Cuban socialism notwithstanding, the government had always confronted that intractable dilemma. After 1961, the problem of motivating workers while promoting social justice and economic development had confounded the Cuban leadership. Three decades later, they were no closer to forging a substitute set of work incentives to sustain socialism in Cuba.

One of the primary targets of the rectification was curbing salaries paid out regardless of work performance and production output. Workers in two key economic sectors—sugar and capital goods—had most often benefited from excessive salaries.[10] As state controls over enterprises tightened, the average salary quickly decreased. Workers in industry and construction suffered salary reductions of about 2 percent. However, although overall average wages in the productive sphere declined about .5 percent, they increased nearly 2 percent in the non-productive sectors. Workers in public administration experienced the highest increase, about 3 percent.[11] Not surprisingly, many workers viewed the rectification as a drive to reduce their salaries. In June 1987, Castro affirmed that the rectification was not primarily about salary reductions. He had also previously noted: "The state does not steal, the state collects for the people. . . . The state is not a capitalist boss."[12] Most workers, however, continued to manifest an immediate *conciencia* of their interests, and many failed to see how the state collected for the people. Consequently, the leadership was forced to reiterate the nominal *raison d'être* of the state in socialism.

Calls for stricter labor discipline also underscored the predicament of state socialism. Over the long run, Decree No. 32 on labor discipline had not proven to be as effective as expected; the unions successfully defended workers against management sanctions. In 1985, for example, somewhat under 1 percent of the labor force appealed discipline sanctions in municipal courts. Of approximately 32,000 appeals, only 38 percent confirmed administrative decisions. Unions tended to argue their cases more effectively than management, and municipal courts ruled in favor of workers in three out of five cases.[13] History had come full circle. During the 1940s, Cuban capitalists had expressed dismay at the 3 : 5 ratio in favor of workers on judicial appeals and had similarly argued that economic growth depended on improving labor discipline and productivity. At the outset of rectification, the Central Committee chastised the unions for their "indolence and tolerance" of indiscipline and payment of excessive salaries.[14] The task for unions was to secure a full-day's work from workers and not to be so preoccupied with end-of-the-year bonuses.[15] In 1988, Castro created a bit of a stir when he noted that highly productive Japanese workers had only six paid vacation days a year while Cubans enjoyed thirty. Shortly thereafter, Castro explained that he did not mean to suggest that vacations be shortened but, rather, that workers "work hard" to improve productivity.[16] Cuban leaders called for the elimination of "paternalism" from an "excessively protective, generous, benevolent, magnanimous" state.[17] They also asserted that extant labor laws undermined labor discipline and economic growth.[18]

The Cuban government was entangled in a web of *conciencia* and efficiency. Indeed, improved labor productivity was critical for economic growth. Salaries incommensurate with work performed undermined productivity as well as *conciencia* of national interests. Upgrading norms and linking salaries to actual output made economic sense. The immediate consequence was the reduction of salaries in the productive sphere, however. As the spirited insistence on *conciencia*, the reduction of salaries, and the ever-increasing shortages took their toll, workers had decreasing incentives to work. The ideal of upholding social justice in the face of austerity resulted in salary increases for the lowest paid workers who were not likely to lead the drive to improve productivity. In contrast, the workers who suffered salary reductions were more likely to contribute to economic development. Motivating workers with more productive potential required linking *conciencia* of their immediate interests with those of the nation. The rectification largely rejected the proposition that material incentives were the primary means to make that link. With the special period, the tensions between efficiency and *conciencia* were further aggravated. Under the conditions of the early 1990s, success at renewing socialism increasingly appeared to be a chimera.

The Politics of Rectification

The politics of rectification also emerged in two stages. Between 1986 and 1989, Cuban leaders emphasized the political importance of the Communist party, which in their view the SDPE had sidelined. With the goal of enterprise profitability as a new "invisible hand," the economic management and planning system had defined largely "economistic" functions for the PCC. Party tasks revolved mostly around the implementation of economic directives. Communists, the rectification claimed, had to exercise political leadership by placing "man"—not profitability—at the center of their work. And, politics meant appealing to higher ideals to motivate people to work, reasons that went beyond satisfying their immediate concerns. Material inducements alone had not led the Cuban people to make the revolution. Commitments to *la patria* and social justice had been more central inspirations. The rectification intended to rescue that élan and incorporate it into the politics of socialism.

The new politics was first evident in the restitution of old forms of labor mobilization. Microbrigades, construction contingents, and volunteer work exemplified the reinforcement of politics to attain economic results. During the early 1970s, microbrigades had been used mainly to build apartment houses. Enterprises formed microbrigades with their own workers and then distributed the apartments in local assemblies. With the establishment of the SDPE, planners and managers had minimized their importance. JUCEPLAN had asserted the Construction Ministry would be more efficient at building more and better quality housing. And, in fact, housing construction expanded from nearly 28,000 units in 1981 to about 42,000 a year during the mid-1980s. In the peak year of 1984, moreover, private construction accounted for 32 percent of the total.[19] Nonetheless, the SDPE had not emphasized socially useful construction like day-care centers and clinics. The microbrigades did *and* mobilized workers to promote collective well-being.

Although construction contingents had purposes similar to those of microbrigades, they mobilized construction workers, the unemployed, and many Angolan veterans. The leadership and the media often hailed contingent workers who normally worked overtime.[20] Work in construction microbrigades and contingents was paid; volunteer work, which the SDPE had also disregarded, was not. In 1987, more than 400,000 people volunteered at least forty hours each to meet various social and economic tasks.[21] After 1990, when the government proclaimed the food plan, thousands of city dwellers went once again to the countryside to do agricultural tasks. Whether paid or voluntary, these forms of work emphasized collective *conciencia* and collective needs. The SDPE had instead enhanced individual interests and private gain.

Conciencia, however, was a problem. Undoubtedly, the revolution had awakened national pride and a sense of empowerment in the Cuban people. Indeed, *conciencia* was in the streets when Ernesto Guevara wrote about the creation of new human beings. The rectification was trying to revitalize that original spirit, which the SDPE had allegedly weakened. The Cuban leadership did not accept that—with or without the SDPE—the enthusiasm of the early years could not possibly have been the same after three decades. Appeals to *conciencia* and mass mobilizations were, nonetheless, the hallmark of the Cuban Revolution, and the rectification resorted to them, but compulsion often appeared to be a more effective lever than *conciencia*.

Stricter labor laws were believed to be the means to attain better labor discipline. Penalties were levied against those who persisted in their "mercantile habits."[22] When many enterprises continued to pay excessive salaries, the government created special groups to enforce the policy on wages.[23] People violated the law so extensively that in 1986 the National Assembly president asserted that all laws would from then on be respected.[24] The rectification had evidently not curbed the misuse and neglect of state resources, and the black market continued to flourish. The party ordered the dismissal of all political cadres, state functionaries, and mass organization leaders who bought stolen merchandise.[25] At the same time, the National Assembly passed a new code that decriminalized many illicit activities and reduced jail sentences for numerous crimes, and a regulation for deleting past sanctions from the dossiers of workers who improved their labor discipline.[26]

The Communist party was central to the rectification process. The SDPE had tended to substitute "mechanisms" for political direction. The party had been more concerned with its own internal affairs than with exercising vanguard leadership. A "new style" was therefore needed to reinforce "contact with the masses." The PCC mandated that provincial and municipal cadres visit enterprises, schools, and neighborhoods to identify problems, provide solutions, and convey explanations. Regular institutional channels often failed to feel the popular pulse. A vanguard party had to listen to ordinary Cubans, establish a dialogue with them, and gain their confidence. As in the early 1970s, the leadership called upon party cadres to "lead," not to "administer."[27]

The first session of the 1986 congress did not, however, dwell on PCC deficiencies. Fidel Castro noted, "With profound satisfaction the party arrives at its third congress stronger, more united and better organized than ever, with ever increasing links to the working class and the rest of the popular masses."[28] Like its assessment of the SDPE, the February self-evaluation did not fully correspond with the view the rectification would emphasize shortly thereafter. In July, the plan of action listed numerous economic and administrative problems that surely had been known but not analyzed in February. As the rectification took shape, the need for a "new style" of work for the party likewise assumed great significance. Castro initiated the rectification; the party had to implement it. In December, the closing session of the congress endorsed the new call to strengthen the party. Castro stressed the centrality of the party, and the origins of rectification highlighted the continued preeminence of charismatic leadership.

The rectification had a populist undercurrent. The party called upon workers to be inspectors at their work centers. The lowest paid workers received salary increases. Austerity measures reduced perquisites for functionaries and cadres. Citizens like a Santiago woman who was fired from her accounting job for denouncing bookkeeping and payroll inconsistencies were exalted in the media. After reviewing her case, the national party reinstated her with back pay and dismissed those who had fired her.[29] Cuban leaders emphasized themes reminiscent of the campaign against bureaucracy during the 1960s. They lambasted the "absolutely reactionary" bureaucrats and technocrats who were incapable of understanding popular needs.[30] Managers tended to be "indifferent, negligent, irresponsible," and somewhat resistant to change.[31] Political and administrative cadres—not the working class—were principally to blame for the state of affairs that had led to the rectification.[32] By the end of 1988, the government had laid off or transferred to productive work more than 6,300 administrative workers and 16,400 administrative cadres.[33]

Nonetheless, the rectification was not confronting another crucial conundrum of socialism: the dilemma of formal democracy. In Cuba and elsewhere, the downfall of capitalism had not resulted in new forms of democracy. Although the revolution had notably expanded popular involvement in public affairs, one-party politics severely constrained the right to dissent. The "power motive," as Alberto Mora had warned in 1965, frequently replaced the "money motive." An overblown bureaucracy was not the only obstacle to maintaining popular support. In the past, the Cuban leadership had repeatedly issued calls for strengthening the party, reinforcing "contact with the masses," and broadening worker participation without sustained success. Political democracy required separation of powers. Democratization in the workplace depended on greater enterprise autonomy, effective material incentives, and worker supervision over labor discipline. More meaningful participation demanded systematic worker input into the appointment and recall of managers. Ultimately, democracy meant political contestation, and the Cuban leadership was as resistant as ever to the formalities of democratic politics.

The rectification, moreover, was emphasizing the higher callings of *la patria* and the revolution to reaffirm national unity during increasingly adverse times. Claiming legitimacy on the basis of their history, Cuban leaders contended that they were the repository of the national honor and the popular well-being. They seemed implicitly to be arguing that only history could pass judgment upon their exercise of power. Consequently, the Communist party could not be subjected to the supervision of elections. The party in "contact with the masses" was its own check and balance. The Central Committee had, for example, approved the *Granma* article on the Santiago accountant. Castro had unequivocally stated:

> Nobody should imagine that somebody on his own can write an article judging the state, the party, the laws, but especially the party. We want broad information, but . . . nobody can assume the prerogative of judging the party. This should be very clear. Ours is not a liberal-bourgeois regime.[34]

The rectification relied primarily on moral principles to safeguard the exercise of power. Charismatic leadership had initiated the new policies, and cadres were the primary link between the party and the masses. Institutional channels frequently obstructed the "correct" execution of political orientations. Only when the Santiago woman had appealed to the national party had she obtained a review of her case. All other intermediate structures had failed her. Castro had a special advisory commission to keep fully abreast of domestic developments. The

Central Committee and the Council of Ministers were not sufficient. The rectification, therefore, was remiss in providing guidelines for governing Cuba after Fidel Castro. Moreover, only charismatic authority rendered the discourse of moral exhortations, revolutionary sacrifice, and radical nationalism possible. Whatever the international circumstances of the late 1980s and early 1990s, the Cuban government would have had to face the prospect of the transition from the historic leadership. The end of the cold war imparted that eventual transition with ominous prospects. For the first time since the rout of the U.S.-sponsored Bay of Pigs invasion in 1961, the downfall of the government was a real possibility. The rectification was, nonetheless, resorting to the politics of revolution at a time when the viability of socialism was quite uncertain.

In 1989, a new stage began in the politics of rectification. Domestic developments first underscored the urgency of revising the status quo. The summer of 1989 notoriously highlighted the pitfalls of charismatic authority and the weaknesses of Cuban institutions, especially the Communist party. In June, the government arrested Division General Arnaldo Ochoa, Hero of the Republic of Cuba, veteran of the wars in Ethiopia and Angola, and commander-designate of the Western Army; the Interior Ministry's Colonel Antonio de la Guardia; and twelve other high-ranking military and security officers. Their arrests became a crisis of scandalous proportions. The fourteen—individuals of high prestige, singular valor, and impeccable credentials—faced accusations of drug trafficking and endangering national security. Their actions lent credence to long-standing U.S. charges and had given the United States cause to take preemptive strikes against Cuba. Because world opinion justly repudiated the drug trade, the Cuban government would have had little recourse against the United States.

After a military court found the defendants guilty, Ochoa, de la Guardia, and their two top aides were put to death before a firing squad. The others received jail sentences ranging from 10 to 30 years.[35] The party also discharged Interior Minister José Abrantes and the top commanders of the security establishment. Widespread dismissals in the security apparatus, other ministries, and the network of agencies dealing with tourism and foreign trade were likewise reported.[36] Abrantes was eventually arrested, brought to trial, and sentenced to a 20-year prison term for dereliction of duty. He was, however, absolved from having known about or partaken in drug trafficking.[37] Even though Ochoa was an army general, Armed Forces Minister Raúl Castro and other high-level military officers did not suffer the same treatment.

The trials of 1989 might have had a political script. General Ochoa might have been involved in an incipient movement for reform within the armed forces, especially among Angolan veterans. Had that been the case and the drug scandal not eliminated him, Ochoa might have been well positioned to advance their cause as the commander-designate of the powerful Western Army. His actions and those of Antonio de la Guardia and the other officers might have been authorized, or at least known and overlooked, at the highest levels of the Cuban government. Whether or not there was a conspiracy or high-ranking complicity was, in important ways, secondary. The crisis underscored a crucial dimension of Cuban politics. The exercise of charismatic authority depended on a special relationship with the "masses" as well as steadfast elite loyalty. Although after 1970 institutions had increasingly mediated between the government and the citizenry, they had not overridden the dynamics of charismatic authority.

The summer of 1989 bared the central predicament of the leadership of Fidel Castro. The need for elite allegiance resulted in the tolerance of wide-ranging behavior among high-ranking officials. The government had long faced the problem that many among the leadership lived well beyond the means of the average citizen.[38] Privileged lives in the face of an ideology of equality and an economy of austerity had undoubtedly eroded state legitimacy. The scandals of 1989, moreover, had originated in the armed forces and the Interior Ministry, the two pillars of the government. In addition, the history of graft and malfeasance in Cuba before 1959 rendered the issue of corruption particularly sensitive. Nonetheless, corruption had to reach truly brazen proportions for action to be taken against the culprits. Whether or not he knew about the Ochoa–de la Guardia undertakings, Fidel Castro was ultimately responsible for the events of 1989. The crisis was primarily about the patterns of governance that had failed to monitor such behavior and about a political system whose institutions did not function, or functioned weakly.

Internationally, the year 1989 likewise underscored the imperative of political change. Crisis in the Soviet Union and the demise of Eastern European socialism signaled the end of the socialist camp and belied the premise of irreversability of socialism. One-party politics and command economies had run their course: socialist democracy and sustained economic development had proven to be beyond their means. The people— workers, peasants, the intelligentsia—repudiated state socialism. The collapse of communism in Eastern Europe confronted the Cuban government with the international shambles of the political model that had consolidated the revolution. Emphasizing the roots of the revolution and socialism in Cuban history, the rectification now aimed to renew the political system and renovate popular adherence to the current leadership.

In a September 1989 editorial in *Granma*, the party called for an institutional *perfeccionamiento* (improvement), denied that the recent trials had constituted a crisis, and declared cases like those of Ochoa and the Interior Ministry officers to be exceptional.[39] *Granma* argued that the political system had dealt efficaciously with these unusual cases. Nonetheless, the party implicitly recognized that the summer of 1989 had strained political legitimacy. "We must say it clearly: what has just happened has revealed a series of flaws which, in one way or another, involve all the institutions of the Revolution." The party became the central component of the *perfeccionamiento*. The PCC congress—initially slated for 1990, then scheduled for the first semester of 1991, and finally held October 10–15, 1991—and its preparatory process offered the leadership the opportunity to renew legitimacy and regain credibility. In February 1990, the Central Committee issued a call for the congress and, a month later, Raúl Castro delivered the *llamamiento* (convocation).[40]

The Cuban Communist Party and the Future of Cuban Socialism

The document calling the congress of the Communist party constituted the Cuban leadership's answer to the double crisis of 1989. The *llamamiento* reaffirmed the national origins of revolution and socialism in Cuba, the benefits and dignity that the past three decades had brought the Cuban people, and the enduring legitimacy of the mandate of 1959. Claiming once again the mantle of the independence struggles that the Ten Years' War had begun in 1868, the PCC declared that only the present leadership had the wherewithal to uphold that heritage. The Baraguá Protest of 1878, when Antonio Maceo had refused to lay down his arms against the Spanish army, was the symbol par excellence of Cuban resistance against foreign impositions and one that the party wholeheartedly appropriated.

Contrary to the prevailing international consensus, the Cuban leadership emphasized the paramount importance of revolutionary ideology, social property, and economic planning. The *llamamiento* reiterated the primacy of the PCC: "The Cuban Communist Party is now and always the party of the Revolution, the party of socialism, and the party of the Cuban nation." That assertion notwithstanding, the document emphasized the need to allow the expression of different currents of opinion, avoid the "unrealistic eagerness" to achieve unanimity, and refrain from political discrimination, especially against religious believers. These differences, however, were possible only within what the Cuban leadership understood to be the national unity necessary to safeguard national independence, socialism, and the revolutionary heritage. Consequently, the party moved away from the politics of sectorial interests of the 1970s and early 1980s. Held in January and March of 1990, the CTC and FMC congesses gave evidence of the new political turn.

The CTC and the FMC in the Rectification Process

Since 1986, when the rectification began to curtail the economic management and planning system, the CTC had experienced a relative retrenchment. The SDPE had allowed a partial coincidence of interests between managers and workers. The unions had also proven to be effective in defending workers against management-imposed sanctions. Under the rectification, the Cuban leadership criticized the CTC for abetting practices that were contrary to national interests and socialist objectives. The actual *conciencia* of the working class was undermining historical visions, and the unions were not combating the many manifestations of widening economism. The challenge was to promote the correct *conciencia*, and thus the party charged the CTC with reining in economistic tendencies, fostering labor discipline, improving enterprise efficiency, and curtailing the persistent problem of stolen goods and off-schedule services that fueled the black market. In 1989, the party removed General Secretary Roberto Veiga and appointed Pedro Ross to preside over the preparations for the CTC congress in January 1990.[41] Thus, the PCC once again violated CTC statutes whereby the recall of leaders was a union prerogative.[42]

Within the limits of vanguard politics, the relative institutionalization of the 1970s and early 1980s had strengthened the unions to act on behalf of workers and opened the possibility of greater rank-and-file involvement in enterprise affairs. After 1986, the rectification curbed the SDPE, and the question of the proper role of trade unions resurfaced. The CTC congress addressed it under circumstances that were already inauspicious. Unlike the congresses of the 1970s and early 1980s, the congress in 1990 did not have the "space" for union activity that the SDPE had defined. The compass of the rectification was more difficult to adjust for the daily conduct of union affairs. Promoting correct *conciencia* was a more abstract task than securing larger wage funds, bonuses, and other material rewards. In addition, an economic recession that would soon become a near-collapse further constrained the unions. National prerogatives took precedence over sectorial interests. Symbolic of the predominant ambience was the attendance of congress delegates in militia uniforms.

The CTC report to the congress highlighted the central charge of developing *conciencia* among workers. New forms of labor mobilization and renewed efforts to enhance worker participation in enterprise management received particular emphasis. Delegates again considered matters of labor discipline. The CTC insisted on greater worker control over the process of labor justice and the need for some degree of collective supervision of enterprise management. The primary concern, however, was to promote *conciencia* of the legitimate interests of workers within the purview of national priorities. Although the congress emphasized moral and collective incentives, the labor leadership did not dismiss the significance of individual, material rewards for high-quality, efficient work. Like the party, the CTC advocated a "new style" of leadership and special attention to ordinary workers and their needs. The congress underscored the importance of keeping "contact" with the rank and file and echoed critiques of bureaucratic methods similar to those of the rectification.[43]

Like the radical experiment and the institutionalization, the rectification failed to acknowledge that workers and unions might have cause to challenge the government and the national leadership. However dissimilar the two earlier periods had been, both had, nonetheless, found support in then-prevailing domestic and international conditions. The rectification stood on much less certain grounds. Moreover, the themes raised at the CTC congress of 1990 were not novel. Since 1961, the role of unions in a centrally planned economy, the tensions between central supervision and local autonomy, the relationship between workers and managers, and the problems of wages, discipline, and efficiency had been constant elements. Most important, the CTC congress still grappled with the question of how to make workers behave like the owners of the national wealth that socialism proclaimed them to be. The Cuban leadership continued to assume workers would manifest new work attitudes and national perspectives once they acquired the right *conciencia*. Neither the radical experiment nor the institutionalization had promoted that elusive *conciencia*. The early 1990s were did not augur success to the rectification.

The FMC congress in March likewise differed from previous gatherings of the organization during the 1970s and early 1980s. The congress document echoed many of the themes that the party and the federation had promoted since the mid-1970s. The FMC highlighted persistent obstacles to gender equality such as the double burden of work and home, continued discrimination in job promotions, and the enduring lag in promoting women to leadership positions. The federation especially emphasized the problems of working women and the need to work more closely with the CTC. The FMC document underscored the new stage the organization was entering because of membership changes. During the early 1960s, *federadas* were largely housewives without previous political experience. Three decades later, FMC members covered the full spectrum of Cuban women—professionals, students, housewives, workers, retirees—with a long record of political and social activities.[44]

Nonetheless, the FMC congress did not fully discuss the document. More so than the earlier CTC gathering, the FMC meeting focused on the national situation. In February, the electoral defeat of the Sandinistas had stunned the Cuban leadership. Although in his address Fidel Castro reiterated the importance of equality between women and men, the emphasis was on the reaffirmation of socialism, the imperative of national defense, and the pursuit of new economic strategies. Although the priority the congress placed on the need to recruit young women was one of the few inklings to the crisis of purpose confronting the organization, there was little else that anticipated the criticisms that the party assemblies in the summer of 1990 would express.[45] Neither could an observer of the congress have anticipated that FMC President Vilma Espín would be excluded from the Politburo in October 1991. The downturn of the FMC from functioning relatively well until the mid-1980s to having its existence questioned and its president marginalized during the early 1990s has yet to be fully documented.[46]

The Fourth Party Congress

Mass discussions of the *llamamiento* for the party congress constituted the first step in what Cuban leaders hoped would be a process of renovating popular consent. In April, the PCC suspended the first round. The discussions were being carried out in rote fashion with an "unrealistic eagerness to achieve unanimity." Calling for a "conscious and active participation" and a culture of debate, the party reconvened the assemblies in the summer.[47] In June, however, *Granma* published a Politburo note establishing the debate boundaries: the one-party system, the socialist economy, and, implicitly, the leadership of Fidel Castro.[48] The PCC said the precongress process was one of "meaningful consultation" and "political clarification."[49] In the second round, the citizenry took up the summons to debate and express critical opinions like never before since 1959. Nonetheless, the party could take solace that the substantial turnout also represented a manifestation of popular confidence. The assemblies overwhelmingly favored reforms within socialism: reopening peasant markets, legalizing self-employment, and holding direct elections to the National Assembly of Popular Power.[50]

The congress preliminaries also included an internal PCC process of elections and restructuring. Between January and April 1990, militants voted for their local leaders using secret ballots rather than raising their hands.[51] Throughout 1990 and early 1991, municipal and provincial assemblies similarly selected their leaderships. The intraparty elections turned over up to 50 percent of the municipal leaders and two of the fourteen provincial secretaries and that of the Isle of Youth.[52] The turnover, however, did not result from competitive elections but, rather, from the failure of party commissions to renominate the incumbents. Moreover, the electoral procedure of a single list of candidates was not modified. The significant change was listing 20 to 25 percent more candidates than those required to fill the positions. Militants marked two *X*s beside the name of their choice for first secretary.

In October 1990, the PCC announced the restructuring of party committees, cut back the number of professional cadres, eliminated ten of the nineteen Central Committee departments, and reduced the Secretariat from ten to five members.[53] Cuban leaders had always equated bureaucratic retrenchment with more effective governance, and they were thus following their long-held beliefs in streamlining the party apparatus. However limited, elections also had an unprecedented importance in the precongress preparations: local party committees elected one-third of the delegates and suggested candidates for the Central Committee. The party hierarchy selected the remaining two-thirds and drew up the final list of CC nominees.[54]

Cuban leaders had never entertained democracy to mean the right to dissent from the ideals of national independence and social justice as they understood them. They claimed the legacy of Cuban history as their mantle and denied the possibility of alternate ways of defending *la patria* and promoting social justice. They had always considered central planning and vanguard politics to be the bulwark of state power. The *llamamiento* discussions and the October 1991 congress were never charged—as the Politburo note well made clear—with questioning the one-party system, the socialist economy, and Fidel Castro. In January 1991, the party leadership concluded:

> There was unanimous, unqualified support for the Revolution, for the Party, for its policies. . . . What is significant and far reaching about this process is its positive balance sheet in favor of the Party, the Revolution, and the First Secretary of our Party, comrade Fidel. It was . . . a plebiscite in favor of socialism with the right to debate.[55]

The uncontested unity of revolutionary politics was therefore constraining the scope of the changes that Cuban society so eminently required.

On October 10, 1991, the Communist party convened its fourth congress in Santiago de Cuba. Less than two months had passed since the failed coup against Mikhail Gorbachev. During the uncertain days of late August, the Cuban government had issued a subdued statement about noninterference in Soviet internal affairs.[56] Nonetheless, the prospect of a successful coup surely raised the hopes of Cuban leaders for a modest stabilization of trade with the Soviet Union. Had the coup succeeded, they would have, more importantly, recovered a politico-ideological ally. When the PCC congress convened, the Cuban leadership knew that all hopes of reinstating the status quo ante in the rapidly decomposing Soviet Union had definitively vanished. Cuba was decidedly alone. The congress emphasized the historic roots of the revolution and socialism in Cuba and declared the PCC to be "sole party of the Cuban nation, *martiano*, Marxist, and Leninist."[57] On the 123d anniversary of the start of the Ten Years' War, the PCC met near the Moncada Barracks and the Sierra Maestra. The parallels between the Cubans who had taken up arms against Spain and those who now defied all odds to preserve the nation abounded. A substantial contingent of military and security officers, including Armed Forces Minister Raúl Castro and Interior Minister Abelardo Colomé, were absent from the initial sessions: they were safeguarding *la patria* from their command posts so that the congress could take place. About one-third of the delegates were veterans from internationalist missions. Nearly 1,700 Cuban communists were charged with saving *"la patria,* the revolution, and socialism."[58]

Unlike previous congresses, this congress did not feature Fidel Castro reading a main report that reviewed past economic performance, the mass organizations, the state administration, the party itself, and other matters. Instead, Castro made a long opening speech concentrating on ideological issues, the disappearance of the socialist countries, the disintegration of the Soviet Communist Party, the uniqueness of Cuban socialism, and the place of foreign investment in national development. In an effort to explain an increasingly pressing economic situation, he detailed the significant shortfalls in Soviet deliveries of oil, raw materials, and food products during the first nine months of 1991.[59] However unintentionally, Castro also rendered a thorough inventory of Cuban dependence on the Soviet Union: its extent seriously undermined the claim to independence so central to the revolutionary heritage. During the 1970s and early 1980s, Cuban leaders had accepted the terms of the relationship with the Soviet Union without incorporating into their development strategies a reduction of dependence. They had forsaken the ideal of a more balanced relationship with the world economy that had inspired their efforts during the 1960s. Undoubtedly, the Cuban government had faced exceptionally contrary odds because of the United States. Nonetheless, the shortfall in Soviet deliveries laboriously tallied to the party congress was also an indictment: the Cuban leadership had failed to achieve a more balanced economy to sustain the nation.

Congress delegates considered resolutions on PCC statutes, the party program, foreign policy, Popular Power assemblies, and the granting of special powers to the Central Committee. Their substance did not depart from the course that the rectification had taken, especially after 1989. The resolutions, nonetheless, contained changes and ideas of some note. The party eliminated the Secretariat and alternate-member status to the Central Committee. Discrimination against religious believers was banned. Direct elections to the National Assembly and provincial assemblies of Popular Power were approved. The PCC recognized the need to increase popular participation—"in organized and constructive ways"—to attain "the necessary consensus."[60] In spite of popular demand, the congress did not approve the reopening of peasant markets but consented to limited forms of self-employment. The party emphasized political and economic integration into Latin America and promised Latin American investors preferential treatment under the new policy of attracting foreign capital. Finally, the congress conferred upon the Central Committee "special powers" in the event of situations endangering *"la patria,* the revolution, and socialism."[61] The declaration of a state of emergency was, thus, a possibility the Communist party evidently contemplated.

Nonetheless, the most significant outcome of the PCC congress was continued elite turnover: 67.1 percent of the 225 members of the Central Committee were either newly elected or promoted to full membership (see Table 7.1). The shares of the state administration, the mass organizations, and the military declined. Most pointed were the losses of the state administration and the mass organizations: the first symbolized the demise of the institutionalization; the second, the relative decline of the historic organizations of the Cuban Revolution. Additional military decreases constituted continued recognition of the imperatives of civilian domestic concerns and the need to give those who represented them at least symbolic presence in the leadership. Among the traditional categories, only the PCC increased its share of CC membership, a good indication of the importance the rectification accorded politics over administration. The number of individuals in other activities increased prominently from 13.7 percent to 35.1 percent. Being an ordinary Cuban in production, research institutes, and social services was now the most important avenue to the Central Committee. Civilian and domestic imperatives received recognition individually; in 1986, the party had recognized their importance institutionally. Current CC members were also more representative of the provinces: in 1986, 17.1 percent had worked in the provinces; in 1991, 34.2 percent of the membership did so. The average age of the 1991 CC was 47 years; that of the 1986 CC had been 52. The Politburo expanded to twenty-five members and excluded historical figures like FMC President Vilma Espín and Culture Minister Armando Hart.[62]

The October 1991 party meeting confirmed the capacity of the Cuban leadership to renovate elites, maintain consensus among them, and prevent significant ruling group divisions. The controversies over the pace and extent of reform that undoubtedly existed were not evident at

Table 7.1. Central Committee Composition, Cuba, 1986, 1991 (in percentages)

	1986 (N = 146)	1991 (N = 225)
PCC	24.7	28.9
State	26.0	14.2
Military	17.8	13.8
Mass organizations	13.0	5.8
Other	13.7	35.1
Advisory Commission	4.8	2.2
Provinces	17.1	34.2

Sources: Granma, Supplement, February 8, 1986, and Pablo Alfonso, Los fieles de Castro (Miami: Ediciones Cambio, 1991).

the congress. No individual or group seemed to have the convictions or the resources to challenge Fidel Castro; most in the Cuban elite probably agreed that they still needed him to remain in power, even if the rectification and the special period were not the policies they would advocate without him. Cuban leaders, moreover, had no self-doubts about their right to govern Cuba. "We are the only ones and there is no alternative," Fidel Castro told the party congress.[63] Nonetheless, a central question was how long Cuban elites would remain united behind Fidel Castro and a course that seemed to be compromising their viability once he passed from the scene. For the time being, they were, however, united and had yet to face a credible opposition with a significant following. The character of elites and the absence of a tenable alternative appeared to indicate that a transition to a new government would be slower than the mounting crisis suggested.

At the end of 1991, the hardest question confronting the Cuban government was how long the Cuban people would continue to consent—out of conviction, fear, or passivity—to be governed in the same, or almost the same, ways as in the past. The party congress forcefully insisted on one-party politics and central planning. The Cuban Revolution had mobilized extraordinary popular support and had offered the Cuban government a long-standing source of legitimacy. During the 1980s and early 1990s, domestic and international conditions had partially eroded that legitimacy and undermined the viability of Cuban socialism. Indeed, popular discontent was widespread.[64] Young people were especially resisting the old ways. Intellectuals were more freely expressing criticism of official cultural policy. Ordinary Cubans were increasingly unreceptive to the appeals to *la patria*, the revolution, and socialism. More than ever, their *conciencia* focused on the excruciating difficulties of their daily lives. Even if they could not or did not want to imagine an alternative, too many citizens no longer had hope in the future. Although the will, energy, and passion of the Cuban people had long supported the government, their fear, apathy, and sense of impotence now played an important part in its stability.

Conclusion

By the early 1990s, Cuban socialism was becoming increasingly untenable. The international conditions that had buttressed it rapidly disappeared. After 1989, the Cuban economy no longer could count upon a network of trade, credit, and aid with the Soviet Union and Eastern Europe. The end of the cold war had also weakened Cuban national security. In a bipolar world, the Soviet Union had provided a partial shield against the United States, and the Cuban government had been able to formulate an activist foreign policy that likewise earned it a modicum of security. In contrast, the post–cold war era was not receptive to the tenets of one-party politics and central planning that Cuban leaders so defiantly defended. The world that had allowed the Cuban Revolution to consolidate and the Cuban government to sustain socialism had come to an end.

The new order did not, however, abate U.S. hostility. The United States had never looked kindly upon Cuban independence nor had it ever had normal relations with Cuba. Except for the Carter administration, U.S. governments had never articulated an enlightened policy toward Cuba—before or after 1959. With the demise of the Soviet Union and the end of armed conflicts in Central America, Cuba could no longer be construed as a threat to the U.S. national interest. Even so, Washington reinforced the three-decade-old embargo with the Cuban Democracy Act of 1992 and precluded any form of rapprochement with the Cuban government until it, in effect, capitulated. The United States was wagering that a harder line would finally bring the downfall of Fidel Castro and that his successor would uproot the legacy of the social revolution. Indeed, the words of John Quincy Adams, written in 1823, echoed with contemporary resonance:

> If an apple severed by the tempest from its native tree cannot choose but fall to the ground, Cuba, forcibly disjoined from its own unnatural connection with Spain, and incapable of self support, can gravitate only towards the North American Union.[1]

Although the United States had had a historic concern with Cuban political stability, its efforts to promote it before 1959 had been singularly unsuccessful. Under the Platt Amendment, the U.S. government twice intervened militarily, and otherwise oversaw many aspects of Cuban domestic affairs. Mediated Cuban sovereignty contributed to the revolutionary upheavals of the 1930s. During the 1950s, the United States supported Fulgencio Batista because he promised order after the mounting chaos of the late 1940s and early 1950s. When, twice, the moderate opposition sought to negotiate an electoral transition, the Eisenhower administration refrained from pressuring General Batista to accede. Had Batista negotiated, the weight of Fidel Castro, the Rebel Army, and the July 26th Movement in the opposition movement would not have been as pronounced.

In the early 1990s, the United States was similarly undermining the future prospects for political stability in Cuba. Confrontation had not moderated Cuban leaders in the past, nor was it mitigating their intransigence at this time. On the contrary. By giving Fidel Castro cause to appeal to the patriotism of millions of Cubans, the United States was continuing to fuel radical nationalism. Because of U.S. hostility, the Cuban leadership had a credible pretext for refusing to implement meaningful changes, and consequently the probabilities of peaceful transformation were diminishing. The United States, moreover, was not likely to be as tolerant of a Tiananmen Square–like massacre in Havana as it had been in Beijing. Were the U.S. military to mediate a governmental transition, the long-term political stability of Cuba would inexorably suffer.

The domestic circumstances of Cuban socialism had also changed. Cuba was no longer in revolution, and socialism had long dulled popular effervescence. Ernesto Guevara had argued that the challenge was to incorporate into daily life the *conciencia* of the confrontations with the *clases económicas* and the United States and the "spirits" of Playa Girón, the Literacy Campaign, and the Missile Crisis. Although Cuban leaders had attempted to forge a *sui generis* socialism during the radical experiment, the outcome had been economic chaos and popular demoralization. After the debacle of 1970, they had no choice but to turn to the models of state socialism to organize the economy and institutionalize the political system. Although the policies of the 1970s and early 1980s had brought relative success, Cuban leaders repudiated them when some of their consequences contravened the social equality and the sense of justice that they deemed necessary to maintain national unity against the United States. The rectification and the special period in peacetime were efforts to rescue the legacy of the revolution amid the domestic and international crisis of socialism.

The probabilities of success were dim. The economy needed foreign capital, international credits, and a new network of trade that did not appear to be forthcoming at the levels required to support recovery. Cuban workers still had to be motivated to produce efficiently. For more than three decades and using different models of economic organization, Cuban socialism had not solved one of the major dilemmas of state socialism. That, amid the conditions of the early 1990s, the government would find the right formula to motivate labor and renovate socialism seemed highly implausible. Moreover, the Cuban leadership was already confronting the gradual breakdown of socialism. The second economy was entrenched and expanding. With deepening crisis in the state sector, ordinary Cubans were more frequently participating in the black market as buyers, producers, and sellers. Without efficiency and growth, state ownership of the means of production was not sufficient to uphold a socialist economy. In addition, foreign investment, though limited, was creating "capitalist islands" in the Cuban economy that likewise underscored the inefficiency of the socialist sector.

Similarly, a new *conciencia* had generally not developed. Ordinary Cubans had often been capable of extraordinary heroism, courage, and dedication. Guevara's examples of Playa Girón, the Literacy Campaign, and the Missile Crisis were authentic. So were the mobilizations for volunteer work during the 1960s. That these were not completely voluntary did not diminish the fact that millions of citizens participated in them with commitment and in the hope of building a better future. During the 1970s and early 1980s, more than 300,000 Cubans served in Angola and Ethiopia with great valor and efficacy. Thousands more also rendered their services in medicine, education, and other civilian professions throughout the Third World. Thus, during extraordinary times and circumstances, many Cuban citizens—sometimes millions—performed well and quite selflessly. Incorporating that *conciencia* into the daily life of state socialism had, however, turned out to be a chimera.

Political factors also intervened in the untenable future of Cuban socialism. Cuba was no longer in revolution, but Cuban politics retained the logic of Fidel-*patria*-revolution. Fidel Castro believed that he had the right to govern Cuba because of the social revolution and the achievements of socialism. The memory of the overwhelming popular support that he had mustered and maintained for so long nurtured this belief in the face of the adverse current of domestic and international events. Fidel Castro and other Cuban leaders were, moreover, convinced that the Cuban nation could exist only under the terms of the past three decades, and that, therefore, only their rule could safeguard *la patria*. Under charismatic authority, Cuban politics was unlikely to achieve the level of institutionalization necessary to promote a peaceful transformation.

The anticipated transition from the leadership of Fidel Castro was a crucial dimension of the political crisis that confronted the Cuban government. Had Castro died in the late 1970s or early 1980s, the Communist party might well have engineered it without momentous consequences. Whatever the domestic and international conditions, the party would have had to face the prospect of governing on its own sometime in the 1990s or after the turn of the century. Under actual circumstances, the crisis was graver because charismatic authority was itself in question and Fidel Castro very much alive. Moreover, the long-run viability of the party—especially when Castro passed from the scene—was uncertain.

How much strength the Communist party had independent of Fidel Castro was one of the principal questions besetting Cuban politics after 1959. One measure of its institutional strength would be its ability to handle the transition under the inauspicious conditions of the 1990s. If the party were able to do so peacefully, even if it lost power while remaining a political force in Cuban society, the conclusion about its strength would be positive. An experience like that of the Sandinistas in Nicaragua would prove the organizational viability of the Cuban Communist Party. If the party were not able to lead a peaceful transition and the result was a violent catastrophe, its inability to negotiate and survive would be a testament to its institutional weakness. If the outcome in Cuba were one of complete disavowal of the revolutionary legacy, it would partially be the consequence of the frailties of the party.

The Cuban Revolution had long been immensely popular. In 1961, when the revolution disallowed capitalism and representative democracy, the overwhelming majority of *el pueblo cubano* supported the government against the invaders at the Bay of Pigs. Hardships notwithstanding, most Cubans remained committed to revolution during the 1960s. Moreover, after the debacle of 1970, Cuban leaders succeeded in engaging the citizenry in the process of institutionalization. On July 26, 1970, Fidel Castro had indeed been able to stand in the Plaza of the Revolution and, however rhetorically, say he would resign if the people so desired. At the time, few Cubans could or wanted to conceive an alternative, and the government retained the support of the majority. The social revolution was but a decade away and the will, energy, and passion that it had generated could still be tapped.

Rejecting the *politiquería* of the past, the revolution had embraced the model of a vanguard party and mass organizations to establish a new political authority. During the early 1960s, institutional politics had not, however, complemented the imperatives of charismatic authority and mass mobilizations. The radical experiment sought to revive the revolutionary "spirit" but it failed dramatically. During the 1970s and early 1980s, the process of institutionalization created the settings for the involvement of the citizenry in issues of immediate relevance. Nonetheless, involvement—whether through mass mobilizations, trade unions, the FMC, or local assemblies of Popular Power—was not tantamount to meaningful participation. Moreover, at no time after 1959 did the Cuban government recognize the right to dissent from the revolution, socialism, and the leadership of Fidel Castro.

During the early 1990s, the dynamic of Fidel-*patria*-revolution and its links to the Communist party as a formula for governance was weakening. The achievements of the social revolution had long legitimated the Cuban government. Self-affirmation against the United States had consolidated a nationalist consensus. The fall of communism in the Soviet Union and Eastern Europe, however, cast a pall on the politics of vanguard parties. In addition, the post-1959 social transformations— which extended and advanced the relative modernity of the old Cuba— were rendering obsolete the forms of political authority established after 1959. New generations of healthier, better-educated, more urban Cubans were demanding the right to express their creativity, their interests, and their political diversity. The Communist party, however, was falling considerably short of their democratic expectations. Indeed, the nation urgently needed to hear their voices and deliberate their visions in order to find new ways of defending its sovereignty and upholding social justice into the new century.

The greater civilian presence in the Central Committee aside, Cuban leaders responded to adversity by emphasizing military imperatives in the conduct of politics. They governed as if still presiding over a social revolution and chose not to hear the words of José Martí to Máximo Gómez: "You do not found a nation, General, the way you command an army camp." Because they continued to give praetorian answers to the special kind of low-intensity warfare that the United States waged against them, Cuban leaders were coming perilously close to commanding the nation like an army camp. And, in so doing, Fidel Castro and the Communist party were failing to meet the historic task of safeguarding the legacy of the revolution.

In 1992, the Cuban government had two occasions to signal a disposition for meaningful reform. In July, the National Assembly discussed a series of constitutional changes dealing with foreign investment in the Cuban economy, freedom of religion and guarantees against discrimination, the functioning of Popular Power assemblies, and other topics. The

Assembly completely revised 42 and updated 34 of the 141 articles in the constitution. Article 141 stipulated that a referendum be held in the event of a total constitutional reform, or a change in the powers of the National Assembly or the Council of the State. Although it was debatable that the quantity and type of the changes legally required a referendum, the government might have called one for political reasons but did not.[2] In October, the National Assembly considered a new electoral law to guide the direct election of the delegates to the National Assembly and the provincial assemblies of Popular Power. Allowing for some diversity among the delegates might have been an indication of willingness to engage in a national dialogue. Although elections contested by multiple parties were probably an unrealistic expectation, the nomination of citizens with different viewpoints was not. In March, a high-ranking party official had, indeed, indicated that dissidents could aspire to office.[3] However, the National Assembly approved a process that virtually precluded the nomination of anyone outside official circles. Electoral commissions sanctioned the candidates to the municipal assemblies, and the municipal delegates in turn selected half of the provincial and national candidates from among their ranks; the trade unions and other mass organizations nominated the other half.[4] In the municipal elections, at least two candidates were nominated from each district. In the national and provincial elections, the total number of candidates equaled the number of delegates allotted to each district, and therefore they ran unopposed. At all levels, the law banned traditional campaigning; the only platform was *la patria*, the revolution, and socialism. Thus, the Cuban leadership could not tolerate the prospect of even a small number of opposition delegates in the assemblies of Popular Power.

Held in December 1992 and February 1993, the elections constituted the last step in the process of *perfeccionamiento* of the Cuban political system that began in 1989. In December, more than 97 percent of the electorate voted for nearly 14,000 municipal assembly delegates. Various sources estimated that about a third of the citizenry cast invalid ballots (blank or defaced in some way); the government admitted to less than 15 percent.[5] Because the elections were not meaningfully competitive, the choice to invalidate the ballot was tantamount to an anti-government vote. Whatever the extent of invalidation, Cuban leaders prepared for the election of the 589 deputies to the National Assembly and the 1,190 delegates to the provincial assemblies in February as they had not in December. The government created an atmosphere of greater supervision, control, and pressure. Before February 24, it updated voter registration lists, increased the number of polling places, and sent representatives from the Committees for the Defense of the Revolution to visit every home to instruct the citizenry on the new voting procedures. Even though candidates had no challengers, citizens could vote selectively from the complete list. They were urged to vote the entire slate and thus express their support for *la patria,* the revolution, and socialism. Fidel Castro, who ran in a district in Santiago de Cuba, campaigned as if, indeed, he had an opponent. His only opponent was, in fact, the U.S.-sponsored Radio Martí which, prior to the December and February elections, called upon the Cuban electorate to cast invalid ballots or vote selectively.

On February 24, more than 99 percent of the electorate went to the polls. Official sources claimed that 88.5 percent voted the straight ticket and only 7.2 percent invalidated the ballot.[6] In his district, Castro won with more than 99 percent of the vote. Unofficial sources placed the proportion of invalid ballots at 10 to 20 percent and the percentage of citizens voting selectively at 30 percent.[7] Citing the rather incredible results, the government claimed "an Olympic victory." Nonetheless, under the domestic and international conditions of the early 1990s, how indicative of the popular will could uncompetitive elections be? Would official candidates have fared as well had they had to defend the government record? Moreover, the real measure of success was not the officially proclaimed near-unanimity but whether the government had bolstered its legitimacy in the eyes of the Cuban people. With the process of *perfeccionamiento* concluded, Cuban leaders continued to govern on the basis of uncontested unity around Fidel-*patria*-revolution. And therein lay a weakness that contradicted the self-assurance of their official rhetoric.

Nonetheless, the Cuban government retained an undetermined level of popular suport. For many citizens, breaking with the government meant breaking with their lives: they had grown up or were young adults during the 1960s, when the social revolution engulfed Cuban society, and they had committed themselves to the new Cuba. Many others—particularly poor and nonwhite Cubans—remembered their plight before the revolution and feared a postsocialist Cuba that would disregard their welfare. Moreover, a majority of the citizenry—even if unsupportive of the government—would still rally to its side in the event of U.S. intervention. The revolution had, indeed, consolidated a widespread sense of nationalism even if socialism had not developed the economy to sustain it.

But there was also the Cuba of *la doble moral* (duplicity) that the party had recognized as an obstacle in the efforts to renovate Cuban socialism during the preparations for the October 1991 congress. Uncontested unity around Fidel-*patria*-revolution had imposed regimentation upon public discourse and exacted a high toll on the expression of dissent. The citizenry—out of conviction, fear, or passivity—had accepted this dynamic and, therefore, governing with a single party under charismatic authority had been, if not democratic, plausible. Domestic and international conditions, however, were undercutting the politics of revolution. Fidel Castro could no longer stand before the nation and offer to resign: a majority of *el pueblo cubano* might have welcomed his departure. To varying degrees, growing numbers of citizens were living in a second society: acquiescing in public and dissenting in private. And the second society was as detrimental to the politics of socialism as the second economy to its economics.

At the beginning of the 1990s, the Cuban government confronted a political crisis. International circumstances had certainly aggravated its dimensions; its domestic origins were, however, just as important. The social revolution was history, and the government faced the consequences of socialism and the ways in which it had exercised political power. Like Gerardo Machado in the 1930s and Fulgencio Batista in the 1950s, Fidel Castro was extraordinarily reluctant to compromise his rule.[8] Unlike Machado and Batista and to his historic credit, Fidel the revolutionary had consolidated a Cuba of greater equality and sovereignty. Castro the caudillo, however, was undermining the legacy of the revolution. Unlike the 1930s and the 1950s, an opposition movement with a credible program to sway *el pueblo cubano* had not developed. Because they knew the threat of such a movement from their experience against Batista, Cuban leaders were redoubling their emphasis on elite unity. They were insisting that their rule was the only safeguard for *la patria* and were doing whatever was necessary to prevent the mounting popular dissaffection from developing an organized expression. If they persisted in their intransigence, the outcome was likely to be the abandonment of the legacy of the Cuban Revolution.

Notes

Introduction

1. Among Cuban historians, see Jorge Ibarra, *Ideología mambisa* (Havana: Instituto del Libro, 1972); Oscar Pino-Santos, *El asalto a Cuba por la oligarquía financiera yanqui* (Havana: Casa de las Américas, 1973); Ramón de Armas, *La revolución pospuesta* (Havana: Editorial de Ciencias Sociales, 1975); and Francisco López Segrera, *Raíces históricas de la revolución cubana (1868–1959)* (Havana: Ediciones Unión, 1980). The Cuban leadership has also espoused the "one-hundred-years-of-struggle" thesis. See . . . *Porque en Cuba sólo ha habido una revolución* (Havana: Departamento de Orientación Revolucionaria, 1975) for speeches by Fidel Castro and Armando Hart. Among U.S. historians, see Ramón Eduardo Ruiz, *Cuba: The Making of a Revolution* (Amherst: University of Massachusetts Press, 1968); Sheldon B. Liss, *The Roots of Revolution: Radical Thought in Cuba* (Lincoln: University of Nebraska Press, 1987); and Louis A. Pérez, Jr., *Cuba: Between Reform and Revolution* (New York: Oxford University Press, 1988).

2. Linearity and teleology often plague coherent and suggestive interpretations of history. Nonetheless, because the revolution has emphasized the "logic" of Cuban history, alternative assessments need to respond—intellectually and politically—to its challenges. The revolution turned history into a factor of politics that contending perspectives cannot ignore—but generally have.

3. For analyses of Martí and the PRC, see Jorge Ibarra, *José Martí dirigente político e ideario revolucionario* (Havana: Editorial de Ciencias Sociales, 1980); John M. Kirk, *José Martí: Mentor of the Cuban Nation* (Tampa: University Presses

of Florida, 1983); and Gerald E. Poyo, *"With All, and for the Good of All": The Emergence of Popular Nationalism in the Cuban Communities in the United States, 1848–1898* (Durham: Duke University Press, 1989).

4. The following are representative of the vast post-1959 literature: Richard Fagen, *The Transformation of Political Culture in Cuba* (Stanford: Stanford University Press, 1969); James O'Connor, *The Origins of Socialism in Cuba* (Ithaca: Cornell University Press, 1970); Maurice Zeitlin, *Revolutionary Politics and the Cuban Working Class* (New York: Harper & Row, 1970); Edward González, *Cuba Under Castro: The Limits of Charisma* (Boston: Houghton Mifflin, 1974); Carmelo Mesa-Lago, *Cuba in the 1970s: Pragmatism and Institutionaliza-. tion* (Albuquerque: University of New Mexico Press, 1974), and *The Economy of Socialist Cuba* (Albuquerque: University of New Mexico Press, 1981); Jorge I. Domínguez, *Cuba: Order and Revolution* (Cambridge: Harvard University Press, 1978); Claes Brundenius, *Revolutionary Cuba: The Challenge of Economic Growth with Equity* (Boulder: Westview Press, 1984; and Andrew Zimbalist and Claes Brundenius, *The Cuban Economy: Measurement and Analysis of Socialist Performance* (Baltimore: Johns Hopkins University Press, 1989).

5. See Theodore Draper, *Castro's Revolution: Myths and Realities* (New York: Praeger, 1962), and *Castroism: Theory and Practice* (New York: Praeger, 1965); Grupo Cubano de Investigaciones Económicas, *Un estudio sobre Cuba* (Miami: University of Miami Press, 1963); and Andrés Suárez, *Cuba: Castroism and Communism, 1959–1966* (Cambridge: MIT Press, 1967).

6. O'Connor is the foremost proponent of the stagnation thesis. Most sympathetic accounts of the revolution, however, subscribe to it.

7. Leo Huberman and Paul Sweezy, *Cuba: Anatomy of a Revolution* (New York: Monthly Review Press, 1961); and Eric Wolf, *Peasant Wars of the Twentieth Century* (New York: Harper & Row, 1969).

8. C. Wright Mills, *Listen Yankee!* (New York: Ballantine Books, 1960).

9. Boris Goldenberg, *The Cuban Revolution and Latin America* (New York: Praeger, 1965).

10. Draper, *Castroism;* and Samuel Farber, *Revolution and Reaction in Cuba, 1933–1960* (Middletown, CT: Wesleyan University Press, 1976).

11. Carlos Rafael Rodríguez, *Cuba en el tránsito al socialismo* (Havana: Editora Política, 1979).

12. Zeitlin.

13. Sidney W. Mintz, "The Rural Proletariat and the Problem of Proletarian Consciousness," *Journal of Peasant Studies* (April 1974): 291–325.

14. Two important exceptions are Hugh Thomas, *Cuba: The Pursuit of Freedom* (New York: Harper & Row, 1971); and Domínguez.

15. I am using *emerge* in the structural sense that Theda Skocpol, *States and Social Revolutions: A Comparative Analysis of France, Russia, and China* (Cambridge: Cambridge University Press, 1979), used it. This study, however, also emphasizes political and ideological factors in the making of the Cuban Revolution.

16. Quite obviously, I am positioning myself in the controversy about the relative weight of structural and political factors in the processes of revolution. Although not wholly subscribing to the emphasis on structural explanations of her early work, I am indebted to Theda Skocpol for her impressive work on social revolutions. I am equally indebted to Charles Tilly and his work on collec-

tive action, state formation, and social transformations. See, especially, *From Mobilization to Revolution* (Reading, MA: Addison-Wesley, 1978), and "Does Modernization Breed Revolution?" in Jack A. Goldstone, ed., *Revolutions: Theoretical, Comparative, and Historical Studies* (San Diego: Harcourt Brace Jovanovitch, 1986), pp. 47–57. The literature on revolutions is enormous and varied. Two recent and quite useful reviews are Rod Aya, *Rethinking Revolutions and Collective Violence: Studies on Concept, Theory, and Method* (Amsterdam: Het Spinhuis, 1990); and Michael S. Kimmel, *Revolution: A Sociological Interpretation* (Philadelphia: Temple University Press, 1990). Although I have refrained from engaging the literature on revolutions, I have used its analytical categories in writing *The Cuban Revolution*.

17. *Clases económicas* and *clases populares* were commonly used terms in the old Cuba for the bourgeoisie and the popular sectors, respectively. That the bourgeoisie used the self-referent of "economic classes" raises the question of whether they considered the working class to be noneconomic. That "popular sectors" rather than the "proletariat" was in such common usage is explained by widespread unemployment and underemployment.

18. André Gunder Frank, *Capitalism and Underdevelopment in Latin America* (New York: Monthly Review Press, 1967); Immanuel Wallerstein, *The Modern World System*, vol. 1 (New York: Academic Press, 1974); and Charles Tilly, *Big Structures, Large Processes, Huge Comparisons* (New York: Russell Sage Foundation, 1984).

19. Cristóbal Kay, *Latin American Theories of Development and Underdevelopment* (London: Routledge, 1989).

20. Peter B. Evans and John D. Stephens, "Development and the World Economy," in Neil J. Smelser, ed., *Handbook of Sociology* (Newbury Park, CA: Sage, 1988), pp. 739–773.

21. Susan Eckstein, "State and Market Dynamics in Castro's Cuba," in Peter Evans, Dietrich Rueschemeyer, and Evelyne Huber Stephens, eds., *States versus Markets in the World-System* (Beverly Hills: Sage, 1985), pp. 217–245, is an important exception.

22. Fernando Henrique Cardoso and Enzo Faletto, *Dependency and Development in Latin America* (Berkeley: University of California Press, 1979).

23. Alejandro Portes and John Walton, *Labor, Class and the International System* (Orlando: Academic Press, 1981).

24. Alain de Janvry, *The Agrarian Question and Reformism in Latin America* (Baltimore: Johns Hopkins University Press, 1981).

25. Charles Bergquist, *Labor in Latin America* (Stanford: Stanford University Press, 1986).

26. Ruth Berins Collier and David Collier, *Shaping the Political Arena: Critical Junctures, the Labor Movement, and Regime Dynamics in Latin America* (Princeton: Princeton University Press, 1991).

27. Torcuato S. Di Tella, *Latin American Politics: A Theoretical Framework* (Austin: University of Texas Press, 1990).

28. John Sheahan, *Patterns of Development in Latin America: Poverty, Repression, and Economic Strategy* (Princeton: Princeton University Press, 1987).

29. Peter Evans, *Dependent Development: The Alliance of Multinational, State, and Local Capital in Brazil* (Princeton: Princeton University Press, 1979); David Becker, *The New Bourgeoisie and the Limits of Dependency* (Princeton:

Princeton University Press, 1983); Gary Gereffi, *The Pharmaceutical Industry and Dependency in the Third World* (Princeton: Princeton University Press, 1983); Maurice Zeitlin, *The Civil Wars in Chile* (Princeton: Princeton University Press, 1983); Maurice Zeitlin and Richard Earl Ratcliff, *Landlords and Capitalists: The Dominant Class of Chile* (Princeton: Princeton University Press, 1988); Mauricio Font, *Coffee, Contention, and Change in the Making of Modern Brazil* (Oxford and Cambridge: Basil Blackwell, 1990); and Carmenza Gallo, *Taxes and State Power: Political Instability in Bolivia, 1900–1950* (Philadelphia: Temple University Press, 1991).

30. In Marifeli Pérez-Stable, "The Field of Cuban Studies," *Latin American Research Review* 1 (1991): 239–250, I argue that the study of Cuba should be called "Cuban studies," not "Cubanology," and those who study Cuba "Cubanists," not "Cubanologists." "Cubanology" and "Cubanologists" situate Cuba within the old Soviet and Eastern European studies and derail the study of Cuba from the Latin American context.

Chapter 1

1. Thomas, pp. 1562–1563.

2. Ibid., p. 1563. By 1925, the Cuban share of world sugar production (cane and beet) had increased more than eightfold. Using the 1-million-ton harvest of 1894 as the base, production increased fivefold, and the share of world production 60 percent.

3. Estimated from Oscar Pino-Santos, *El asalto a Cuba por la oligarquía financiera yanqui* (Havana: Casa de las Américas, 1973), p. 93; and Byron White, *Azúcar amargo: un estudio de la economía cubana* (Havana: Publicaciones Cultural, 1954), p. 27.

4. Thomas, pp. 1563–1564.

5. I calculated per capita sugar output using Thomas, pp. 1563–1564; and Comité Estatal de Estadísticas, *Anuario Estadístico de Cuba, 1986*, p. 59.

6. Computed from Julián Alienes Urosa, *Características fundamentales de la economía cubana* (Havana: Banco Nacional, 1950), p. 52; Banco Nacional de Cuba, *Memoria, 1949–1950* (Havana: Editorial Lex, 1951), p. 70; and Banco Nacional de Cuba, *Memoria 1958–1959* (Havana: Editorial Lex, 1960), pp. 95, 154.

7. Grupo Cubano de Investigaciones Económicas; Raúl Cepero Bonilla, *Política azucarera (1952–1958)* (Mexico: Editora Futuro, 1958); and Arnaldo Silva León, *Cuba y el mercado internacional azucarero* (Havana: Editorial de Ciencias Sociales, 1975).

8. Oscar Zanetti, "El comercio exterior de la república neocolonial," in *Anuario de estudios cubanos: la república neocolonial*, vol. 1 (Havana: Editorial de Ciencias Sociales, 1975), pp. 76–78.

9. Computed from Zanetti, p. 119.

10. José Antonio Guerra, "Necesidad de organizar sobre bases permanentes y nacionales el estudio y determinación de la política comercial internacional de Cuba," in Cámara de Comercio de la República de Cuba and Asociación Nacional de Industriales de Cuba, eds., *Conferencia para el progreso de la economía nacional* (Havana, 1949), p. 212; and White, p. 76.

11. Zanetti, pp. 71–72.

12. Gustavo Gutiérrez, *El desarrollo económico de Cuba* (Havana: Junta Nacional de Economía, 1952), p. 76.

13. "El desarrollo económico de Cuba," *Revista del Banco Nacional* 2 (March 1956): 273–276.

14. White, p. 43.

15. Ministerio de Agricultura, *Memoria del Censo Agrícola Nacional* (Havana: P. Fernández, 1951), pp. 286–296; Gutiérrez, p. 23; and Brundenius, *Revolutionary Cuba,* pp. 146–147.

16. Carmelo Mesa-Lago, *The Labor Force, Employment, Unemployment and Underemployment in Cuba: 1899–1970* (Beverly Hills: Sage, 1972), p. 23 for sugar-sector employment, and p. 29 for GNP share.

17. White, p. 150.

18. Louis A. Pérez, Jr., *Cuba Under the Platt Amendment, 1902–1934* (Pittsburgh: University of Pittsburgh Press, 1986), pp. 62–72, 76–77. The Rionda quotation, p. 72.

19. Zanetti, pp. 93–94, 100–101.

20. Pérez, *Cuba Under the Platt Amendment,* pp. 230–231.

21. "La reforma arancelaria," *Revista del Banco Nacional* 1 (enero 1958): 19.

22. Gutiérrez, p. 135.

23. "El desarrollo industrial de Cuba," *Cuba Socialista* 6 (abril de 1966): 135.

24. Pérez, *Cuba Under the Platt Amendment,* p. 230.

25. Leland H. Jenks, *Nuestra colonia de Cuba* (Buenos Aires: Editorial Palestra, 1959), p. 167; and Pino-Santos, p. 76.

26. Jorge I. Domínguez, "Seeking Permission to Build a Nation: Cuban Nationalism and U.S. Response Under the First Machado Presidency," *Cuban Studies/Estudios Cubanos* 16 (1986): 33–48.

27. Ramiro Guerra, *Azúcar y población en las Antillas* (Havana: Editorial de Ciencias Sociales, 1976), is the classic work on sugar-industry reform.

28. Luis Machado, "Necesidad de adoptar una política de comercio exterior," in Gustavo Gutiérrez, ed., *El problema económico de Cuba* (Havana: Molina, 1931), pp. 35–80, proposed a policy of tariff wars, and leading reformers Ramiro Guerra and Gustavo Gutiérrez denounced his proposal.

29. Gutiérrez, *El problema económico,* pp. 97–108.

30. Gustavo Gutiérrez, "Necesidad de adoptar una política exterior," in Gutiérrez, *El problema económico,* p. 24.

31. Luis Machado, *La isla de corcho: ensayo de economía cubana* (Havana: Mazo, Caso, 1936), p. 28.

32. Jenks, pp. 58–63.

33. Zanetti, pp. 94, 101.

34. Alienes Urosa, p. 52. Per capita income was estimated using 1937 prices.

35. Silva León, p. 74; and Jules Robert Benjamin, *The United States and Cuba: Hegemony and Dependent Development, 1880–1934* (Pittsburgh: University of Pittsburgh Press, 1974), pp. 35–38.

36. Zanetti, p. 94, for the rise in Cuban exports to the United States, and p. 101 for U.S. exports to Cuba.

37. International Bank for Reconstruction and Development (IBRD),

Report on Cuba (Baltimore: Johns Hopkins University Press, 1951), p. 733.

38. Ibid., p. 94.

39. Gutiérrez, *El desarrollo económico de Cuba*, p. 135, used 1937 prices.

40. Ibid., pp. 24, 135.

41. Ibid., p. 138; values are given in constant 1937 pesos.

42. Jacinto Torras, *Obras escogidas, 1939–1945* (Havana: Editora Política, 1984), p. 104.

43. *El Mundo*, June 12, 1946, p. 19.

44. Cámara de Comercio, pp. 80, 81–82 for the quotation; p. 71 for the letter from sugar mill owners and cane growers explaining their abstention from the final document; and pp. 79–83 for all recommendations on trade policy.

45. IBRD, *Report*, p. 81; and Cámara de Comercio, pp. 79–81.

46. Ibid., p. 79.

47. Marcos Winocour, *Las clases olvidadas de la revolución cubana* (Barcelona: Editorial Crítica, 1979), pp. 37–64.

48. Consejo Nacional de Economía, *El programa económico de Cuba* (Havana, 1955), pp. 26, 24.

49. Jorge F. Pérez-López, "An Index of Cuban Industrial Output, 1930–1958," in James W. Wilkie and Kenneth Riddle, eds., *Quantitative Latin American Studies: Methods and Findings—Statistical Abstract of Latin America* (Los Angeles: University of California, Los Angeles, Latin American Center Publications, 1977), p. 52.

50. "La reforma arancelaria," p. 19.

51. Gutiérrez, *El desarrollo económico de Cuba*, p. 138; and *Memoria, 1958–1959*, pp. 98, 189.

52. U.S. Department of Commerce, *Investment in Cuba* (Washington, DC, 1955), p. 72.

53. *Memoria, 1958–1959*, p. 96.

54. IBRD, *Report*, p. 73; and U.S. Department of Commerce, pp. 12–15. The latter reported that during the 1950s, Cuban nationals had over $300 million in short-term assets and long-term investments in the United States.

55. *Memoria, 1958–1959*, p. 192.

56. Ismael Zuaznábar, *La economía cubana en la década del 50* (Havana: Editorial de Ciencias Sociales, 1986), p. 111.

57. Computed from Zanetti, pp. 95, 101–102.

58. Computed from Grupo Cubano de Investigaciones Económicas, p. 1245; and Zanetti, p. 115.

59. Asociación Nacional de Industriales de Cuba, *Boletín*, May 15, May 31, June 15, 1955.

60. Asociación Nacional de Industriales de Cuba, *Boletín*, June 15, 1955, p. 13.

61. Pérez, *Cuba Under the Platt Amendment*, p. 265; and Silva León, pp. 31–32.

62. Boris C. Swerling, "Domestic Control of an Export Industry: Cuban Sugar," *Journal of Farm Economics* 3 (August 1951): 346–356.

63. Luis G. Mendoza, *Revista Semanal Azucarera: Selecciones 1935–1945* (Havana: Editorial Lex, 1945), p. 26.

64. *El Mundo*, April–July 1946, July 1947.

65. *El Mundo*, July, August 1947.

66. Torras, pp. 499–502. I am grateful to Louis A. Pérez, Jr., for bringing to my attention the story of rice agriculture in Cuba.

67. República de Cuba, *Informe general del censo de 1943* (Havana: P. Fernández, 1945), p. 419.

68. Torras, p. 499.

69. For rice production and imports, see Grupo Cubano de Investigaciones Económicas, p. 1051, and *Memoria, 1958–1959*, p. 119. For a general overview of the rice story, see Torras, pp. 499–502, 524–528; and Raúl Cepero Bonilla, *Escritos económicos* (Havana: Editorial de Ciencias Sociales, 1971), pp. 258–259, 297–302, 368–375, 384–386. White, p. 49, noted that rice growers were among the most modern entrepreneurs in Cuba. In contrast, U.S. Department of Commerce, p. 5, concluded rather disingenuously that the failure of rice in Cuba was due to faulty farming methods, soil depletion, and irrigation misuse.

70. "La reforma arancelaria," pp. 5–21.

71. U.S. Department of Commerce, pp. 155–162.

72. *Memoria, 1958–1959*, p. 101.

73. U.S. Department of Commerce, p. 11.

74. Carlos Manuel Raggi Ageo, "Contribución al estudio de las clases medias en Cuba," in Theo R. Crevenna, ed., *Materiales para el estudio de la clase media en América Latina*, 6 vols. (Washington, DC: Panamerican Union, 1950–1951), p. 79. In the Cuban context, most of the 38 percent of the labor force whose monthly earnings were 75 pesos or more could be considered middle class. Informe de la Comisión Coordinadora de la Investigación del Empleo, Sub-Empleo y Desempleo, *Resultados de la Encuesta sobre Empleo, Sub-Empleo y Desempleo en Cuba (mayo 1956-abril 1957)* (Havana, January 1958), p. 52.

75. República de Cuba, *Censos de población, viviendas y electoral: informe general (enero 28 de 1953)* (Havana: P. Fernández, 1955), p. 204.

76. U.S. Department of Labor, *Foreign Labor Information: Labor in Cuba* (Washington, DC, 1957), p. 10.

77. *Informe general del censo de 1943*, p. 1043; and *Censos de población, viviendas y electoral*, p. 195. Between 1943 and 1953, the growth of personal services highlighted the increasing demand of the middle sectors. In 1943 and 1953, the total persons in personal services were 73,963 and 178,504, respectively. Their increase was nearly threefold, and their share of the economically active population doubled, from 5 percent to 10 percent.

78. In 1953, unemployment was 8.2 percent during the harvest; only 68.4 percent of the labor force worked full-time throughout the year. In 1956–1957, unemployment averaged 16.4 percent and peaked at 20.7 percent during the dead season. In 1943, off-season unemployment was 21.1 percent; the census did not include underemployment data. I included manufacturing, mining, and construction in industry. In 1953, the partial breakdown of an EAP of 1,972,266 was 327,208 manufacturing; 9,618 mining; 65,292 construction; 818,706 agriculture; 395,904 services; and 232,323 commerce. Computed from *Censos de población, viviendas y electoral*, pp. 153, 169, 176, 195; Informe de la Comisión Coordinadora, pp. 41, 50; and *Informe general del censo de 1943*, p. 1056.

79. Computed from *Censos de población, viviendas y electoral*, pp. 19–21, 153–154, 169, 176. Population centers of 150 people or more with a range of services including electricity, health, legal, and entertainment were considered urban. Fifty-seven percent of the population and 60.0 percent of the labor force

lived in urban areas. The 1931 and 1943 censuses considered "urban" all lo-
calities where addresses were registered with street names. In 1899, 1907, and
1919, the population living in centers of more than 1,000 inhabitants was
counted as "urban." The urban population first outnumbered the rural population in 1931.

80. Agrupación Católica Universitaria, *¿Por qué reforma agraria?* (Havana,
1957), pp. 32, 34–35. The choice of institutions was government, church, masonry, bosses or landlords, trade unions. Seventeen percent responded they
expected *patrones* (bosses) to improve their living conditions; only 7 percent
identified the union.

81. The 1925–1926 school enrollment of 63.0 percent was cited by IBRD,
Report, p. 412. The 1943 and 1953 percentages—40.8 and 51.6—were computed
from *Informe general del censo de 1943,* pp. 482–484; and *Censos de población,
viviendas y electoral,* pp. 32, 99.

82. IBRD, *Report,* p. 406, reported the literacy rates in the census years as
follows: 43.2 percent (1899), 56.6 percent (1907), 61.6 percent (1919), 71.7
percent (1931), and 71.3 percent (1943). For the 1953 rate of 76.4 percent, see
Censos de población, viviendas y electoral, p. 143.

83. I used the data in Grupo Cubano de Investigaciones Económicas and
América en Cifras, 8 vols. (Washington, DC: Panamerican Union, 1960), for all
comparisons between Cuba and the rest of Latin America.

84. Educational data were drawn or computed from *Censos de población,
viviendas y electoral,* pp. 99, 131. The illiteracy rate in urban and rural areas was
11.6 percent and 41.7 percent, respectively. School enrollment among 5- to 14-
year olds was 69.0 percent (urban) and 34.9 percent (rural). Although 60.4
percent of the total population 6 years and older had up to a third-grade level, 55
percent of urban Cubans had beyond that level, and 80 percent of rural below.
Less than 3 percent of university graduates and 5 percent of secondary, vocational, and technical school graduates lived in rural areas. In 1953, there were
52,172 university graduates, 84,716 high school graduates, and 82,374 vocational and technical school graduates in Cuba.

85. Agrupación Católica Universitaria, p. 33.

86. For 1953 life expectancy, see Sergio Díaz-Briquets, *The Health Revolution in Cuba* (Austin: University of Texas Press, 1983), p. 161. For 1952 crude
annual death rate, see *Censos de población, viviendas y electoral,* p. 318. Caution
should be exercised in taking the 6.4 death rate at face value. From the data
reported in the 1953 census, the death rate for the province of Havana was
higher, at 8.3, than for the rest of the country. The Havana figures were probably
higher because of better death registration. Díaz-Briquets, p. 191, ranked cardiovascular diseases, malignant and benign tumors, and diarrhea, gastritis, and
enteritis as the top-three causes of death in 1953. On p. 320 of the 1953 census,
the top-three causes of death in 1951 were cardiovascular diseases, digestive
system diseases, and malignant tumors. See Domínguez, *Cuba: Order and Revolution,* p. 76, for 1953 infant mortality; and Nelson P. Valdés, "Cuba: Social
Rights and Basic Needs" (Paper presented to the Inter-American Commission on
Human Rights, Washington, DC, February 25, 1983), pp. 25–27, for ratios of
doctor and hospital beds to population.

87. IBRD, *Report* p. 441 for the quotation (italics in original), and pp. 441–

443 for its observations on health. The World Bank mission did not do an in-depth study of public health, as it did with the economy.

88. Agrupación Católica Universitaria, pp. 21–28. The survey also reported extensive malnutrition among rural workers (91 percent); most never ate meat, fish, eggs, and bread or drank milk. Fourteen percent had had or had tuberculosis, 13 percent typhoid fever, and 31 percent malaria.

89. Valdés.

90. Agrupación Católica Universitaria, p. 29.

91. *Censos de población, viviendas y electoral*, pp. 208–213, 253. I considered cement walls with stone, cement, or wood floors or wood walls with stone or cement floors to be solid materials.

92. Computed from Ibid., pp. 19, 21, 100, 120, 185–188. Of *habaneros*, 74.3 percent worked full-time and 14.6 percent were underemployed; 65.8 percent of the labor force in the other five provinces had full-time jobs and 17.8 percent were underemployed. For the distribution of plants with more than five hundred workers, see U.S. Department of Commerce, pp. 73–74.

93. Data drawn from Cepero Bonilla, *Escritos económicos*, pp. 416–417; and *Memoria, 1958–1959*, pp. 151–153. Total wages were 716 million pesos in 1952 and 723 million in 1958. Havana's wage bill increased by 84 million pesos; that of Camagüey, Las Villas, and Oriente by 33, 14, and 13 million pesos, respectively. Total wages in Matanzas (35–36 million) and Pinar del Rio (21–22 million) were basically stagnant over the decade. The total wage bill excluded the salaries of sugar agricultural workers and only partially included other agricultural salaries.

94. *Memoria, 1949–1950*, pp. 173–181; *Memoria, 1958–1959*, pp. 140, 151–153; U.S. Department of Commerce, pp. 73–74; *Censos de población, viviendas y electoral*, p. 186; and Cepero Bonilla, *Escritos económicos*, pp. 416–417.

95. Computed from Informe de la Comisión Coordinadora, pp. 58–60; and *Censos de población, viviendas y electoral*, pp. 153, 200–201, 204–205. The 1956–1957 employment survey published the distribution of the labor force by occupation and income (monthly wages of 75 pesos and more or below 75 pesos) but not the provincial breakdown. Using the 1953 census, I approximated the 1956–1957 income distribution by occupation and province.

96. Lino Novás Calvo, "La tragedia de la clase media cubana," *Bohemia Libre* 13 (January 1, 1961): 76.

97. *Memoria, 1958–1959*, p. 190.

98. White, p. 87.

99. Agrupación Católica Universitaria, pp. 6, 63.

100. K. Lynn Stoner, *From the House to the Streets: The Cuban Women's Movement for Legal Reform, 1898–1940* (Durham: Duke University Press, 1991).

101. *El Mundo*, February 11, 1954, p. A9.

102. *Censos de población, viviendas y electoral*, pp. 195, 205.

103. Computed from ibid., pp. 153–154, 176, 195, 202; and Informe de la Comisión Coordinadora, pp. 41, 50.

104. Computed from *Censos de población, viviendas y electoral*, pp. 153–154, 157–158, 200; and Informe de la Comisión Coordinadora, p. 58.

105. Susan Schroeder, *Cuba: A Handbook of Historical Statistics* (Boston: G. K. Hall, 1982), p. 112.

Chapter 2

1. Juan Pérez de la Riva, "Los recursos humanos de Cuba al comenzar el siglo: inmigración, economía y nacionalidad (1899–1906)," in *Anuario de estudios cubanos: la república neocolonial,* vol. 1 (Havana: Editorial de Ciencias Sociales, 1975), pp. 11–44; and Pérez de la Riva, "Cuba y la migración antillana, 1900–1931," in *Anuario de estudios cubanos la república neocolonial,* vol. 2 (Havana: Editorial de Ciencias Sociales, 1979), pp. 5–75.

2. Hobart A. Spalding, Jr., *Organized Labor in Latin America: Historical Case Studies of Workers in Dependent Societies* (New York: New York University Press, 1977).

3. Pérez, *Cuba Under the Platt Amendment* is a compelling analysis of the dynamics of early republican politics.

4. Louis A. Pérez, Jr., *Army Politics in Cuba, 1898–1958* (Pittsburgh: University of Pittsburgh Press, 1976).

5. Pérez, *Cuba Under the Platt Amendment,* pp. 171–181, 189–213.

6. Instituto de Historia del Movimiento Comunista y de la Revolución Socialista de Cuba, *Historia del movimiento obrero cubano, 1865–1958, vol. 1, 1865–1935* (Havana: Editora Política, 1985), pp. 168–177.

7. Louis A. Pérez, Jr., "Aspects of Hegemony: Labor, State, and Capital in Plattist Cuba," *Cuban Studies/Estudios Cubanos* 16 (1986): 49–69.

8. Instituto de Historia del Movimiento Comunista, 1:116–185, 244–248, 257–261.

9. Dana G. Munro, *The United States and the Caribbean Republics, 1921–1933* (Princeton: Princeton University Press, 1974).

10. Justo Carrillo, *Cuba 1933: estudiantes, yanquis y soldados* (Coral Gables: University of Miami, 1985).

11. Lionel Soto, *La revolución del 33,* 3 vols. (Havana: Editorial de Ciencias Sociales, 1977).

12. Foreign Policy Association, *Problemas de la nueva Cuba* (New York: J. J. Little and Ives, 1935); Charles A. Thomson, "The Cuban Revolution: Fall of Machado," *Foreign Policy Reports* 11 (December 18, 1935): 250–260, and "The Cuban Revolution: Reform and Reaction," *Foreign Policy Reports* 11 (January 1, 1936): 262–276.

13. Instituto de Historia del Movimiento Comunista y de la Revolución Socialista de Cuba, *Historia del movimiento obrero cubano, 1865–1958, vol. 2,* (Havana: Editora Política, 1985), pp. 3–57.

14. *Plan Trienal de Cuba* (Havana: Cultural, 1938).

15. Domínguez, *Cuba: Order and Revolution* p. 89.

16. Enrique J. Guiral, "Orientación de la legislación social para el desarrollo de la economía cubana: proyecciones de la reglamentación estatal del trabajo en una economía basada en la empresa libre," in Cámara de Comercio and Asociación Nacional de Industriales de Cuba, eds., *Conferencia para el progreso de la economía nacional* (Havana, 1949), p. 128.

17. Cepero Bonilla, *Escritos económicos* p. 45.

18. Domínguez, *Cuba: Order and Revolution,* pp. 81, 88.

19. Guiral, pp. 124–125.

20. Francisco Fernández Plá, "Política social apropiada para el fomento de

la producción nacional," in *Conferencia para el progreso de la economía nacional*, p. 158; and Guiral, p. 126.

21. Charles A. Page, "The Development of Organized Labor in Cuba" (Ph.D. diss., University of California, Berkeley, 1952), p. 279.

22. *El Mundo*, February–June, 1946.

23. Ibid., May 20, 1947, p. 25.

24. Ibid., May 7, 1946, p. 20.

25. *Diario de la Marina*, February 13, 1945, p. 11.

26. Blas Roca and Lázaro Peña, *La colaboración entre obreros y patronos* (Havana: Ediciones Sociales, 1945).

27. Office of Strategic Services, *The Political Significance and Influence of the Labor Movement in Latin America. A Preliminary Survey. Cuba* (Washington, DC, 1945), pp. 29–30.

28. Torras, pp. 57–68, 103–107, 447–535, 583–586, 696–706.

29. *El Mundo*, February 5, April 9, April 30, May 14, and May 28, 1946, pp. 10, 19, 21, 19, and 19–21, respectively.

30. Office of Strategic Services, p. 16; Page, p. 114.

31. *El Mundo*, February 3, 1946, p. 1.

32. U.S. Department of Commerce, p. 19.

33. Computed from *Memoria, 1949–1950*, pp. 64, 68, and *Memoria, 1958–1959*, pp. 95–96.

34. Guiral, p. 126.

35. Domínguez, *Cuba: Order and Revolution*, pp. 102, 107.

36. Page, p. 129 for the long quotation; p. 112 for the two labor leaders' quotations; and pp. 215–224 for how the *auténtico* labor leadership used the CTC for other purposes.

37. *El Mundo*, March 13, 1946, p. 10.

38. Instituto de Historia del Movimiento Comunista, 2:173.

39. William S. Stokes, "The 'Cuban Revolution' and the Presidential Elections of 1948," *Hispanic American Historical Review* 31 (February 1951): 74.

40. Cámara de Comercio, pp. 47–54.

41. Cepero Bonilla, p. 100.

42. IBRD *Report*, pp. 143, 181.

43. Domínguez, *Cuba: Order and Revolution*, pp. 81, 88–89.

44. IBRD, *Report*, p. 4.

45. Pérez, *Army Politics*, pp. 116–136.

46. For narratives of the anti-Batista movement, see Rolando E. Bonachea and Nelson P. Valdés, eds., *Revolutionary Struggle (1947–1958)*, Vol. 1 of *Selected Works of Fidel Castro* (Cambridge: MIT Press, 1972), pp. 1–119; Carlos Franqui, *Diary of the Cuban Revolution* (New York: Viking Press, 1980); Ramón L. Bonachea and Marta San Martín, *The Cuban Insurrection: 1952–1959* (New Brunswick, NJ: Transaction Books, 1974); and Thomas, pp. 789–1034.

47. Bonachea and Valdés, p. 221.

48. Ibid., pp. 270–271.

49. Domínguez, *Cuba: Order and Revolution*, p. 89.

50. Consejo Nacional de Economía, p. 24.

51. Cepero Bonilla, *Escritos económicos*, 406.

52. Cepero Bonilla, *Política azucarera*, p. 10.

53. Ibid.

54. Page, p. 174.

55. U.S. Department of Commerce, p. 22.

56. *Memoria, 1949–1950,* pp. 64, 68; *Memoria, 1958–1959,* pp. 95–96.

57. Cepero Bonilla, *Escritos ecónomicos,* pp. 267–268.

58. Ibid., pp. 381–382; Instituto de Historia del Movimiento Comunista, 2:288–294; and Evelio Tellería, *Los congresos obreros en Cuba* (Havana: Instituto Cubano del Libro, 1973), pp. 414–415.

59. Cepero Bonilla, *Escritos económicos,* pp. 382; and Instituto de Historia del Movimiento Comunista, 2:290–294.

60. Ibid., 2:319.

61. U.S. Department of State, *Foreign Relations of the United States, Cuba, 1958–1960,* vol. 6 (Washington, DC: Government Printing Office, 1991), pp. 1–116.

62. See Timothy P. Wickham-Crowley, *Guerrillas and Revolution in Latin America: A Comparative Study of Insurgents and Regimes Since 1956* (Princeton: Princeton University Press, 1992), for interesting comparisons with other guerrilla movements in Latin America.

63. See *Foreign Relations of the United States, Cuba,* pp. 158–250.

64. Bonachea and Valdés, pp. 183–186.

65. Ibid., pp. 269–270, 364–367.

66. Regino Boti and Felipe Pazos, *Algunos aspectos del desarrollo económico de Cuba. Tésis del M-26-7* (Havana: Delegación del Gobierno en el Capitolio Nacional, 1959).

67. Bonachea and Valdés, pp. 341–449.

Chapter 3

1. Adolfo Sánchez Rebolledo, ed., *La Revolución Cubana* (Mexico: Ediciones Era, 1972), p. 139.

2. *Revolución,* January 4, 1959, p. 2.

3. Alfred L. Padula, Jr., "The Fall of the Bourgeoisie: Cuba, 1959–1961" (Ph.D. diss. University of New Mexico, 1974), p. 77.

4. Boti and Pazos.

5. *Revolución,* March 2, 1959, p. 19. On page 1 of its March 4 edition, *Revolución* printed the July 26th Movement's notice that no trademarks would be granted to revolutionary symbols.

6. *Revolución,* March 7, 1959, p. 4.

7. Padula.

8. Leonel-Antonio de la Cuesta, ed., *Constituciones cubanas: Desde 1812 hasta nuestros días* (New York: Ediciones Exilio, 1974), p. 260.

9. Thomas, pp. 1215–1218.

10. Juan and Verena Martínez-Alier, *Cuba: economía y sociedad* (Paris: Ruedo Ibérico, 1972), pp. 109–208.

11. *Revolución,* March 7, 1959, p. 1.

12. Ibid., August 19, 1959, p. 1.

13. Padula, pp. 270, 284.

14. *Revolución,* February 11, 1959, p. 14.

15. Ibid., March 16, 1959, pp. 1, 23.

16. Ibid., March 31, 1959, p. 14.

17. Ibid., June 2, 1959, p. 1.

18. See, for example, ibid., March 31, June 8, 10, 15, and September 21, 1959.

19. Ibid., November 19, 1959, p. 3.

20. Ibid., June 4, 1959, p. 2.

21. Ibid., January 30, p. 11; February 6, p. 11; February 19, pp. 1–2; and May 7, 1959, p. 13.

22. Ibid., May 16, 1959, p. 11.

23. Ibid., May 7, p. 13; and May 13, 1959, p. 14.

24. Reynol González, a July 26th Movement labor leader in 1959, told me the meeting never took place because the CTC had a "demagogic attitude" against the industrialists. He believes that an alliance between the union movement and the industrialists might have deterred the communists' control of the CTC and steered the revolution in more moderate directions. Interview in Miami, Florida, January 8, 1990.

25. *Revolución*, May 13, 1959, p. 2.

26. Ibid., January 6, 1960, pp. 1–2.

27. Ibid., April 10, 1959, pp. 1, 2.

28. Thomas, pp. 1074–1075, 1196–1197, 1202–1203.

29. Padula, pp. 137–138.

30. Ibid., pp. 152–158, 225.

31. *Revolución*, June 2, 1959, pp. 1, 16.

32. Ibid., February 10, 1959, p. 2.

33. Ibid., January 16, p. 5; January 27, p. 7; and January 30, 1959, pp. 1, 16.

34. Ibid., January 28, p. 8; and January 31, 1959, p. 8.

35. Ibid., February 6, p. 5; and May 15, 1959, pp. 1–2.

36. Ibid., January 30, 1959, pp. 1, 16.

37. Ibid., January 23, 1959, p. 3.

38. Ibid., January 23, p. 7; January 28, p. 8; January 30, pp. 5, 11; January 31, p. 8; February 4, p. 8; February 5, p. 4; February 6, pp. 1–2; February 7, pp. 1, 16; March 4, p. 5; March 6, p. 6; April 4, p. 4; April 15, p. 1; April 22, p. 5; May 14, p. 6; June 3, p. 4; June 4, p. 6; July 13, pp. 1–2; July 17, p. 5; and July 20, 1959, p. 4.

39. Ibid., January 28, 1959, p. 8.

40. Ibid., January 10, p. 11; January 15, p. 5; and January 19, 1959, p. 5.

41. Ibid., January 30, p. 11; February 6; and July 30, 1959, p. 18.

42. Ibid., January 30, p. 8; March 15, p. 5; April 7, p. 7; May 14, p. 4; and June 4, 1959, p. 4.

43. Ibid., May 15, 1959, p. 5.

44. Ibid., January 31, 1959, p. 8.

45. Ibid., February 27, p. 4; and April 18, 1959, p. 6.

46. Ibid., July 29, 1959, p. 4.

47. Ibid., February 3, p. 5; February 7, p. 2; February 27, p. 4; March 23, p. 4; March 25, p. 4; May 22, p. 4; June 13, p. 4; July 29, p. 5; and August 25, 1959, p. 4.

48. Ibid., February 4, p. 8; February 14, p. 5; March 2, p. 7; March 3, p. 4; March 15, p. 5; and April 18, 1959, p. 4.

49. Ibid., April 25, 1959, p. 4.

50. Ibid., June 13, 1959, p. 5.

51. Ibid., January 24, p. 7; February 6, p. 5; February 12, p. 6; February 16, p. 7; March 2, p. 4; March 13, p. 4; April 1, p. 1; July 29, p. 5; and August 25, 1959, p. 4.

52. Ibid., June 3, 1959, p. 4.

53. Ibid., November 18, 1959, p. 12.

54. Computed from data in *Memoria, 1958–1959*, pp. 151–153.

55. *Revolución*, January 21, p. 16; January 23, p. 3; and June 13, 1959, pp. 1, 16.

56. Héctor Ayala Castro, "Transformaciones de propiedad, control obrero e intervención de empresas en Cuba (1959–1960)," *Economía y Desarrollo* 47 (May–June 1978): 44–69.

57. U.S. Department of Commerce, p. 15.

58. *Revolución*, February 1, p. 6; February 7, p. 2; March 4, pp. 1, 14; March 5, pp. 1, 15; March 18, p. 4; April 6, p. 7; April 16, 1959, p. 4, for examples of enterprises intervened at the request of workers; and Ibid., June 22, 1960, pp. 1–8, for interventions in 1959–1960.

59. Ibid., February 10, 1959, p. 2.

60. Ibid.

61. Ibid., February 11, 1959, p. 11.

62. Reynol González recognized that support for July 26th Movement labor leaders in the 1959 struggles over CTC control depended on the prestige of the revolution and Fidel Castro. Interview in Miami, Florida, January 8, 1990. Roberto Simeón, a labor leader in the oil industry and member of the Revolutionary Student Directorate, likewise told me the July 26th Movement vied for CTC control on the basis of the "olive green." Simeón was referring to the olive-green fatigues of the Rebel Army. Interview in Miami, Florida, December 26, 1988.

63. *Revolución*, January 26, 1959, p. 23.

64. Ibid., January 19, p. 5; January 27, p. 7; January 31, p. 3; February 1, p. 18; February 11, p. 14; March 2, p. 6; April 1, pp. 4–5; and April 2, 1959, p. 1.

65. Ibid., January 8, p. 8; January 15, p. 5; January 17, p. 5; January 21, p. 7; January 30, p. 5; February 3, p. 5; February 5, p. 11; February 9, pp. 1, 5, 23; February 16, p. 4; March 24, p. 1; and April 18, 1959, p. 4.

66. Ibid., February 10, 1959, p. 2.

67. Ibid., April 1, 1959, p. 7.

68. Ibid., April 27, pp. 1, 18; May 4, p. 1; May 5, p. 5; May 7, pp. 1–2, 5; May 13, p. 4; May 18, 1959, pp. 9, 11.

69. Blas Roca, *29 artículos sobre la revolución cubana* (Havana: Tipografía Ideas, 1960), p. 203.

70. Ibid., pp. 201–204.

71. *Revolución*, January 24, p. 13; February 1, p. 21; February 3, p. 8; February 10, 1959, p. 6.

72. Ibid., April 22, p. 5; May 4, p. 7; May 12, p. 4; May 14, p. 6; May 18, pp. 3, 9, 11; and May 23, 1959, p. 7.

73. Ibid., May 19, p. 15; and May 22, 1959, p. 6.

74. Ibid., September 2, 1959, p. 4.

75. Ibid., May 7, pp. 1–2; May 8, pp. 1–2; May 16, pp. 1, 15; July 26, pp. 5–6; July 31, p. 4; August 26, pp. 1, 19; September 2, p. 4; September 3, p. 2; September 7, pp. 1, 4; and September 10, 1959, pp. 1, 17.

76. Ibid., June 16, p. 4; June 17, p. 4; and July 8, 1959, p. 4.

77. Ibid., August 20, 1959, pp. 1, 16.

78. Ibid., July 17, p. 5; July 23, p. 4; July 23, p. 5; August 10, p. 6; and August 14, 1959, p. 4.

79. Ibid., September 14, pp. 1, 6; and September 15, 1959, pp. 1, 2, 5.

80. Carlos Franqui, *Retrato de familia con Fidel* (Barcelona: Editorial Seix Barral, 1981), p. 121; Thomas, p. 1250; and Maurice Zeitlin and Robert Scheer, *Cuba: Tragedy in Our Hemisphere* (New York: Grove Press, 1963), p. 121.

81. *Revolución,* November 7, 1959, pp. 1, 6.

82. Ibid., November 3, 1959, pp. 1, 8, 9.

83. Ibid., November 17, 1959, p. 4.

84. Ibid., November 19, 1959, pp. 1, 2, 15.

85. Ibid., November 21, 1959, p. 4.

86. Ibid., December 3, 1959, pp. 1, 19.

87. Ibid., January 2, p. 8; February 29, 1960, p. 5.

88. Ralph Lee Woodward, Jr., "Urban Labor and Communism: Cuba," *Caribbean Studies* 3 (October 1963): 17–50.

89. *Revolución,* October 5, pp. 1, 13; December 10, p. 1; and December 14, 1960, pp. 1, 8; and Padula, pp. 411–417.

90. *Revolución,* March 11, 1960, p. 4.

91. Ibid., June 4, 1960, pp. 1, 10.

92. Ibid., June 1, 1960, p. 4.

93. Ibid., May 11, 1959, p. 1.

94. Ibid., February 16, 1959, p. 1.

95. Ibid., April 15, p. 1; and September 26, 1959, p. 2.

96. Ibid., April 11, 1959, pp. 1, 14.

97. Ibid., July 6, 1959, pp. 2, 19.

98. Ibid., June 6, 1959, p. 16.

99. Ibid., October 10, 1959, p. 1.

100. Ibid., February 20, 1960, p. 1.

101. Ibid., March 22, 1960, p. 4.

102. Ibid., September 10, 1960, p. 1.

103. Ibid., January 5, 1959, p. 4.

104. Ibid., March 9, 1959, p. 5.

105. Ibid., March 15, p. 3; March 27, p. 13; and May 4, 1959, p. 27.

106. Ibid., April 4, p. 14; June 15, p. 13; July 7, p. 6; July 29, p. 3; and August 1, 1959, p. 4.

107. Ibid., June 4, 1959, p. 6.

108. Ibid., October 13, 1959, p. 6.

109. Ibid., September 14, p. 19; November 20, 1959, p. 5; February 5, p. 3; March 5, 1960, p. 10.

110. Ibid., August 26, 1960, p. 7.

111. Ibid., May 2, 1960, p. 1.

112. Ibid., November 10, 1959, p. 10.

113. Ibid., May 29, 1959, pp. 1, 4.

114. Ibid., July 15, 1959, pp. 1, 19.

115. Ibid., January 2, 1959, p. 3.

116. Sánchez Rebolledo, pp. 139–149.

117. For overviews of Cuba-U.S. relations between 1959 and 1961, see

Thomas, pp. 1255–1271, 1300–1311; Richard E. Welch, Jr., *Response to Revolution: The United States and the Cuban Revolution, 1959–1961* (Chapel Hill: University of North Carolina Press, 1985); and Wayne S. Smith, *The Closest of Enemies* (New York: Norton, 1987), pp. 42–67. *Foreign Relations of the United States, Cuba*, pp. 334–1191 provides a selection of official communications between the U.S. State Department and the U.S. Embassy in Havana on the first two years of the revolutionary government.

118. *Foreign Relations of the United States, Cuba*, p. 395.

119. Ibid., p. 955.

120. *Revolución*, March 29, 1960, p. 1.

121. Ibid., September 1, 1960, pp. 1–2.

122. Ibid., March 4, 1960, p. 13.

123. Ibid., June 9, 1960, p. 1.

Chapter 4

1. U.S. Department of Commerce, p. 7.

2. For example, Regino Boti, "El plan de desarrollo económico de 1962," *Cuba Socialista* 4 (December 1961): 19–32; Edward Boorstein, *The Economic Transformation of Cuba* (New York: Monthly Review Press, 1968); Ernesto Guevara, "The Alliance for Progress," in Rolando E. Bonachea and Nelson P. Valdés, eds., *Ché: Selected Works of Ernesto Guevara* (Cambridge: MIT Press, 1969), pp. 265–296; Juan F. Noyola, *La economía cubana en los primeros años de la revolución y otros ensayos* (Mexico: Siglo XXI Editores, 1978); and Dudley Seers, *Cuba: The Economic and Social Revolution* (Westport, CT: Greenwood Press, 1975).

3. Felipe Pazos, "Lineamientos de una política de desarrollo económico," *Selva Habanera* 579 (August 13, 1955): 1, 6–7, 9–11).

4. Felipe Pazos, *Influencia de la escuela de ciencias económicas en el desarrollo económico del país.* (Santiago de Cuba: Universidad de Oriente, 1955), p. 12.

5. White.

6. U.S. Department of Commerce, p. 5.

7. Boti and Pazos.

8. I am using the term *inclusive development* to mean development from the bottom up, that is, a more equitable distribution of the benefits and costs of development.

9. Noyola, p. 124; Archibald R. M. Ritter, *The Economic Development of Revolutionary Cuba: Strategy and Economic Performance* (New York: Praeger, 1974), pp. 111–116; Brundenius, *Revolutionary Cuba*, pp. 23–25.

10. Ritter, p. 113.

11. Zanetti, pp. 95, 101–102; *Anuario estadístico de Cuba, 1986*, p. 406. Between 1954 and 1957, surpluses were respectively 51.1, 19.1, 17.2, and 35.5 million pesos. The last represented the upsurge in sugar prices during the Suez Canal crisis. Deficits in 1958 and 1959 were 43.6 and 38.8 million pesos. In 1961, Cuba again registered a slight deficit of 12.3 million. The only surplus since then was in 1974.

12. Felipe Pazos, "Comentarios a dos artículos sobre la revolución cubana," *El Trimestre Económico* 29 (January–March, 1962): 9.

13. Ritter, p. 113.

14. *Anuario estadístico de Cuba, 1986,* p. 406.

15. David P. Barkin, "Cuban Agriculture: A Strategy of Economic Development," in David P. Barkin and Nita Rous Manitzas, eds., *Cuba: The Logic of the Revolution* (Andover, MA: Warner Modular Publications, 1973).

16. For analyses of the 1970 sugar drive, see Heinrich Brunner, *Cuban Sugar Policy from 1963 to 1970* (Pittsburgh: University of Pittsburgh Press, 1977); and Sergio Roca, *Cuban Economic Policy and Ideology: The Ten Million Ton Sugar Harvest* (Beverly Hills: Sage, 1976).

17. Cole Blasier, "COMECON in Cuban Development," in Cole Blasier and Carmelo Mesa-Lago, eds., *Cuba in the World* (Pittsburgh: University of Pittsburgh Press, 1979), pp. 225–255.

18. Jorge I. Domínguez, *To Make a World Safe for Revolution: Cuba's Foreign Policy* (Cambridge: Harvard University Press, 1989), pp. 92–99.

19. Rates for imports from the Soviet Union were 1980–1981, $+10.3$ percent; 1981–1982, $+15.6$ percent; 1982–1983, $+13.5$ percent; 1983–1984, $+14.2$ percent; 1984–1985, $+12.5$ percent; 1985–1986, -1.8 percent; 1987, $+1.8$ percent; and 1987–1988, -2.7 percent. Rates for imports from Bulgaria, Czechoslovakia, Hungary, Poland, and the German Democratic Republic were 1980–1981, $+12.0$ percent; 1981–1982, $+21.3$ percent; 1982–1983, $+17.6$ percent; 1983–1984, -1.8 percent; 1984–1985, $+9.7$ percent; 1985–1986, -8.0 percent; 1986–1987, $+4.2$ percent; and 1987–1988, -1.1 percent. Computed from *Anuario estadístico de Cuba, 1986,* pp. 418–419, and Comité Estatal de Estadísticas, *Anuario estadístico de Cuba, 1988* p. 423.

20. Well before the post-1989 crisis, controversy mired the evaluation of Cuban economic performance. Major technical questions arise with respect to the two systems used by market economies (System of National Accounts, or SNA) and by the former centrally planned economies (Material Product Systems, or MPS) to measure total output: Gross Domestic Product (GDP) and Gross Social Product (GSP). Comparing the two measures is troublesome. First, GDP and GSP derive value differently. GSP tends to inflate value because it is based on gross value, e.g., a shoe factory includes the value of leather and other inputs in its total output. GDP is computed on value added, e.g., only value added at the shoe factory is counted in its output. Second, nonproductive services are included in GDP but not in GSP. For the two measures to be comparable, a common denominator in value and nonproductive services is necessary. The use of different methodologies to measure value of output by the Cuban government since 1959 complicates the evaluation of Cuban economic performance. There is no continuous series of macroeconomic indicators. Moreover, while improving over time, the availability and quality of statistical data on the Cuban economy are less than optimal. These issues were debated in *Comparative Economic Studies.* Claes Brundenius and Andrew Zimbalist, "Recent Studies on Cuban Economic Growth: A Review," and Carmelo Mesa-Lago and Jorge F. Pérez-López, "Imbroglios on the Cuban Economy: A Reply to Brundenius and Zimbalist," *Comparative Economic Studies* 27 (Spring 1985): 22–46 and 47–83, respectively; Brundenius and Zimbalist, "Cuban Economic Growth One More Time: A Response to 'Imbroglios,'" *Comparative Economic Studies* 27 (Fall 1985): 115–131; Mesa-Lago and Pérez-López, "The Endless Cuban Economy Saga: A Terminal Rebuttal," and Brundenius and Zimbalist, "Cuban Growth: A Final Word," *Comparative Economic Studies* 27 (Winter 1985): 67–82 and 83–84, respectively.

I culled trends in Cuban economic performance and approximate growth rates from Domínguez, *Cuba: Order and Revolution*, pp. 173–180; Mesa-Lago, *The Economy of Socialist Cuba*, pp. 33–36; Jorge F. Pérez-López, *Measuring Cuban Economic Performance* (Austin: University of Texas Press, 1987), pp. 117–126; Andrew Zimbalist and Claes Brundenius, *The Cuban Economy: Structure and Performance After Three Decades* (Baltimore: Johns Hopkins University Press, 1989); and Jorge Pérez-López and Carmelo Mesa-Lago, "Cuba: Counter Reform Accelerates Crisis," *ASCE Newsletter* (March 1991): 3–5. Growth rates for 1987 and 1988 were published in José Luis Rodríguez, "The Cuban Economy Today," *Cuba Business* 3 (April 1989): 11, 12, and *Granma*, December 26, 1988, p. 4. Estimates for the 1991–1992 decline are from John Paul Rathbone, "Cuba: Current Economic Situation and Short-Term Prospects," *La Sociedad Económica*, Bulletin 20 (1992): 3.

21. Brundenius, *Revolutionary Cuba*, p. 124, and Andrew Zimbalist and Claes Brundenius, "Growth with Equity: Cuban Development in Comparative Perspective" (unpublished paper, n.d.), p. 9.

22. Computed from William LeoGrande, "Cuban Dependency: A Comparison of Pre-Revolutionary and Post-Revolutionary International Economic Relations," *Cuban Studies/Estudios Cubanos* 9 (July 1979): 9, *Anuario estadístico de Cuba, 1986*, p. 422, and *Anuario estadístico de Cuba, 1988*, p. 426. Zimbalist and Brundenius, *The Cuban Economy*, pp. 145–147, argue, however, that by the 1980s the sugar share of total exports at constant 1965 prices had declined to about 60–65 percent.

23. Computed from Thomas, pp. 1563–1564, and *Anuario estadístico de Cuba, 1988*, pp. 57, 243.

24. See James G. Brown, *The International Sugar Industry: Developments and Prospects*, World Bank Staff Commodity Working Papers Number 18 (Washington, DC: World Bank, 1987), for a rather somber overview of sugar prospects. Marcelo Fernández Font, *Cuba y la economía azucarera mundial* (Havana: Instituto Superior de Relaciones Internacionales, 1986), offered a similar assessment of world market prospects while contending that Cuba could continue to expand sugar production because of the markets in the Soviet Union and Eastern Europe.

25. Charles Edquist, *Capitalism, Socialism and Technology* (London: Zed Books, 1985); Claes Brundenius, "Development and Prospects of Capital Goods Production in Revolutionary Cuba," and Carl Henry Feuer, "The Performance of the Cuban Sugar Industry, 1981–1985," in Andrew Zimbalist, ed., *Cuba's Socialist Economy Toward the 1990s* (Boulder: Lynne Rienner, 1987), pp. 97–114, 69–83.

26. LeoGrande, p. 6; and computed from *Anuario estadístico de Cuba, 1986*, p. 100, and *Anuario estadístico de Cuba, 1988*, pp. 99, 410.

27. Sugar shares of industry, agriculture, and GSP computed from *Anuario estadístico de Cuba, 1986*, p. 297, and *Anuario estadístico de Cuba, 1988*, pp. 99, 235, 300. I used the sugar shares of industry and agriculture to estimate the percentage of sugar workers in the industrial and agricultural labor force from *Anuario estadístico de Cuba, 1986*, p. 192, and *Anuario estadístico de Cuba, 1988*, pp. 235, 300.

28. Percentages of industry and agriculture computed from Brundenius,

Revolutionary Cuba, p. 147, and Carmelo Mesa-Lago, *The Labor Force,* p. 23 for sugar sector employment, and p. 29 for GNP share.

29. Computed from *Anuario estadístico de Cuba, 1986,* pp. 423, 425, and *Anuario estadístico de Cuba, 1988,* pp. 427, 429.

30. *Memoria, 1958–1959,* pp. 189, 191.

31. Computed from *Anuario estadístico de Cuba, 1986,* pp. 407–408, 461–463, and *Anuario estadístico de Cuba, 1988,* pp. 410–412, 467; LeoGrande, p. 14.

32. Computed from Grupo Cubano de Investigaciones Económicas, p. 1245, and Zanetti, pp. 78, 115.

33. Computed from *Anuario estadístico de Cuba, 1988,* pp. 57, 410.

34. Computed from *Anuario estadístico de Cuba, 1986,* pp. 415, 419, and *Anuario estadístico de Cuba, 1988,* pp. 419–423.

35. Rodríguez, "The Cuban Economy," p. 12; *Granma Weekly Review,* August 27, 1989, p. 12. Between 1987 and 1988, however, Cuban trade with market economies increased 183.3 million pesos while hard-currency trade deficits declined 114.5 million pesos.

36. Total Cuban debt to the Soviet Union is from an interview Leonid Abalkin, vice-president of the Council of Ministers in the Soviet Union, granted to *Sovietskaia Rossia;* reproduced in *Granma,* May 7, 1990, p. 7. For Cuban debt in hard currency, see *El Nuevo Herald,* May 4, 1991, p. 1.

37. Rodríguez, "The Cuban Economy," p. 11.

38. *El Nuevo Herald,* May 4, 1991, p. 1.

39. Rathbone, p. 2.

40. *CubaINFO Newsletter* 4 (October 27, 1992): 4–5.

41. According to Cuban economist Pedro Monreal, at the end of 1991, the Cuban government had negotiated 30 projects with foreign capital, 130 were in process, and an additional 60 were in the early stages of negotiations. Although there were no official figures on the amounts involved, the optimistic figure of $500 million was often cited. Presentation at the conference "U.S.-Cuba Relations and the Cuban-American Community, A Second Generation Cuban and Cuban-American Perspective," held at Musgrove Plantation, St. Simon's Island, Georgia, November 15–17, 1991.

42. Julie M. Feinsilver, "The Politics of Health in Cuba in the Age of Perestroika: Capitalizing on Biotechnology and Medical Exports," *Cuban Studies/ Estudios Cubanos* 22 (1992): 79–111.

43. Mesa-Lago, *The Labor Force,* p. 40.

44. República de Cuba. *Censo de población y viviendas 1970* (Havana: Editorial Orbe, 1975), p. 427.

45. Mesa-Lago, *The Economy of Socialist Cuba,* p. 122, for 1970–1978, and Brundenius, *Revolutionary Cuba,* p. 135, for 1970–1980.

46. República de Cuba, *Censo de población y viviendas de 1981,* vol. 16, pt. 1 (Havana: Comité Estatal de Estadísticas, 1983), p. cciv. In 1976, the Cuban government reorganized the six provinces into fourteen: Pinar del Río, La Habana, Ciudad de La Habana, Matanzas, Villa Clara, Sancti Spiritus, Cienfuegos, Camagüey, Ciego de Avila, Las Tunas, Holguín, Granma, Santiago de Cuba, and Guantánamo. The new provinces roughly corresponded to the old ones: Pinar del Río, La Habana (La Habana, Ciudad de La Habana), Matanzas, Las Villas (Villa Clara, Sancti Spiritus, Cienfuegos), Camagüey (Camagüey,

Ciego de Avila), and Oriente (Las Tunas, Holguín, Granma, Santiago de Cuba, Guantánamo). I grouped the new provinces according to the old divisions to facilitate approximate comparisons.

47. *Anuario Estadístico de Cuba, 1988,* p. 192.

48. *Censo de población y viviendas, 1970,* p. xvii, and *Censo de población y viviendas de 1981,* p. xl.

49. Computed from *Censo de población y viviendas 1970,* p. 258, and *Censo de población y viviendas de 1981,* vol. 16, pt. 1, p. cxv. In 1970, all centers with 2,000 or more persons and with 500–2,000 residents with no fewer than four of the following—street lights, paved streets, aqueduct, sewerage, medical services, educational center—were considered urban. In 1981, the urban definition was somewhat looser: all centers with at least 2,000 persons; those with 500–2,000 persons with three or more of the listed characteristics, and those with 200–500 persons with all six. In 1953, there were twenty-six centers with more than 20,000 persons representing 61.6 percent of the urban population; in 1970 and 1981, there were thirty-one (71.4 percent) and forty-two (69.8 percent).

50. Computed from *Censo de población y viviendas de 1981,* vol. 16, pt. 2, pp. 149, 181–184, 187. The enrollment of 6- to 24-year-olds was 75.4 percent (urban), 67.2 percent (rural), and 72.6 percent (national).

51. Computed from Ibid., vol. 16, pt. 2, p. 168.

52. Computed from Ibid., vols. 2, 3, and 16, pt. 2, pp. 150–153, 156, 160, 163, 166, 169, 178. The percentage of the population 6 years or older in high school or graduated from high school was 8 percent in urban areas, including Havana, and 2.9 percent in the countryside. The percentage of university students and graduates was 4.4 percent (urban), 6 percent (Havana), and less than 2 percent rural.

53. *Anuario estadístico de Cuba, 1986,* p. 584.

54. Ibid., p. 78.

55. Ibid., p. 624.

56. Domínguez, *Cuba: Order and Revolution,* p. 186; *Granma,* January 7, 1991, p. 4.

57. *Anuario estadístico de Cuba, 1988,* pp. 563–564, 567, 571.

58. See Sarah M. Santana, "The Cuban Health Care System: Responsiveness to Changing Needs and Demands," in Zimbalist, pp. 115–127.

59. See José L. Luzón, "Housing in Socialist Cuba: An Analysis Using Cuban Censuses of Population and Housing," *Cuban Studies* (University of Pittsburgh Press) 18 (1988): 65–83; Jill Hamburg, "Housing Policy in Revolutionary Cuba," in Rachel Bratt, Chester Hartman, and Ann Meyerson, eds., *Critical Perspectives on Housing* (Philadelphia: Temple University Press, 1986), pp. 586–624; Sergio Roca, "Housing in Socialist Cuba," in Oktay Ural, ed., *Housing: Planning, Financing, Construction,* vol. 1 (New York: Pergamon Press, 1979), pp. 62–74.

60. *Censo de población y viviendas de 1981,* vol. 16, pt. 2, pp. 430–432. I classified as "solid construction" all housing with masonry walls.

61. Ibid., pp. 418–424.

62. Ibid., pp. 406–429.

63. See the section on "Completed Housing per Provinces" in Comité Estatal de Estadísticas, *Anuario estadístico de Cuba* for the years 1982–1988. The figures are not fully comparable because in some years the *Anuario* reports

completed housing regardless of condition and in others only those with a habitable certificate. The trends, however, are revealing. The Oriente provincial groups appeared to be losing their share of total completed housing while the Havana provinces were gaining. The trend might have been due to the relative decline of private construction, which seemed more prevalent in Oriente, and the relative priority the state accorded to housing construction in Havana after 1986.

64. *Censo de población y viviendas de 1981*, vol. 16, pt. 2, pp. 438–450.

65. Ibid., pp. 451–453.

66. Sergio Díaz-Briquets, "Regional Differences in Development and Living Standards in Revolutionary Cuba," *Cuban Studies/Estudios Cubanos* 18 (1988): 45–63.

67. Computed from *Anuario estadístico de Cuba, 1988*, p. 197. After 1985, there seemed to be a trend toward diminishing the gaps between Pinar del Río and the Oriente provinces and the national average and reducing the slightly more privileged position of the Havana provinces.

68. Computed from ibid., p. 223.

69. Computed from ibid., p. 279, and *Anuario Estadístico de Cuba, 1982*, p. 191. The value of construction is given at current prices.

70. Computed from *Anuario estadístico de Cuba, 1986*, p. 204.

71. Ernesto Guevara, "The Meaning of Socialist Planning," in Bertram Silverman, ed., *Man and Socialism in Cuba: The Great Debate* (New York: Atheneum Press, 1971), p. 101.

72. For a compilation of the main exchanges in the Great Debate, see Silverman.

73. Karl Marx, *A Contribution to the Critique of Political Economy* (New York: International Publishers, 1981), p. 20.

74. Ernesto Ché Guevara, "El socialismo y el hombre en Cuba," in *Escritos y Discursos* (Havana: Editorial de Ciencias Sociales, 1977) 8:259.

75. See Andrew Zimbalist and Susan Eckstein, "Patterns of Cuban Development: The First Twenty-Five Years," in Zimbalist, pp. 7–24; Andrew Zimbalist, "Incentives and Planning in Cuba," *Latin American Research Review* 24 (1989): 65–93; Carmelo Mesa-Lago, "The Cuban Economy in the 1980s: The Return of Ideology," in Sergio G. Roca, ed., *Socialist Cuba: Past Interpretations and Future Challenges* (Boulder: Westview Press, 1988), pp. 59–100.

Chapter 5

1. Domínguez, *Cuba: Order and Revolution*, pp. 208, 262.

2. Fagen, pp. 47, 50.

3. Guevara, "El socialismo y el hombre en Cuba," pp. 253–272, quotation at p. 256.

4. Suárez.

5. Ibarra, José Martí; José A. Tabares del Real, *Guiteras* (Havana: Editorial de Ciencias Sociales, 1973).

6. Domínguez, *Cuba: Order and Revolution*, p. 321.

7. Fidel Castro, "Discurso del Primer Ministro en el Comité Provincial de Matanzas," *Cuba Socialista* 2 (May 1962): 11.

8. Ibid., pp. 1–27.

9. Domínguez, *Cuba: Order and Revolution*, p. 311. PSP members constituted 40 percent of the ORI National Directorate and 23 percent of the new Central Committee.

10. Ernesto Ché Guevara, "Discusión colectiva: decisión y responsabilidad únicas," in *Escritos y discursos* (Havana: Editorial de Ciencias Sociales, 1977), 5:200.

11. Ernesto Ché Guevara, "Discurso en la Convención Nacional de los Consejos Técnicos Asesores," in *Escritos y discursos*, 5:38.

12. Guevara, "Discusión colectiva."

13. *Revolución*, September 7, 1962, p. 1.

14. Ernesto Ché Guevara, "Discurso clausura del Consejo Nacional de la CTC, 15 de abril de 1962," in *Escritos y discursos* (Havana: Editorial de Ciencias Sociales, 1977), 6:133.

15. Tellería, p. 508.

16. Ibid., p. 494.

17. *Revolución*, November 27, 1961, pp. 3–4.

18. Quoted in Roberto E. Hernández and Carmelo Mesa-Lago, "Labor Organization and Wages," in Carmelo Mesa-Lago, ed., *Revolutionary Change in Cuba* (Pittsburgh: University of Pittsburgh Press, 1971), p. 220.

19. Adolfo Gilly, "Inside the Cuban Revolution," *Monthly Review* 16 (October 1964): 13–20; Carmelo Mesa-Lago, *The Labor Sector and Socialist Distribution* (New York: Praeger, 1968); and Joan Robinson, "Cuba: 1965," *Monthly Review* 18 (February 1966): 10–18.

20. Gilly, p. 17.

21. Cuban Economic Research Project, *Labor Conditions in Communist Cuba* (Miani: University of Miami Press, 1963), p. 101.

22. Gilly, pp. 18–19.

23. *Revolución*, September 3, 1962, p. 5.

24. Ernesto Ché Guevara, "Discurso a la clase obrera, 14 de junio de 1960," *Escritos y discursos* (Havana: Editorial de Ciencias Sociales, 1977), 4:131–132.

25. Guevara, "Discusión colectiva," p. 196.

26. *Revolución*, February 2, 1963, p. 3.

27. Quoted in Hernández and Mesa-Lago, p. 220.

28. *Revolución*, October 27, 1964, pp. 7–10.

29. Guevara, "Discurso clausura del Consejo Nacional de la CTC," p. 129.

30. *Revolución*, September 7, 1962, p. 1.

31. Zeitlin, *Revolutionary Politics*, quotations at pp. 284, 293. In the summer of 1962, Zeitlin interviewed a random sample of 210 industrial workers in twenty-one plants scattered throughout Cuba. This paragraph is based on his findings.

32. Guevara, "El socialismo y el hombre en Cuba," p. 254.

33. Alberto Mora, "On Certain Problems of Building Socialism," in Bertram Silverman, ed., *Man and Socialism in Cuba: The Great Debate* (New York: Atheneum, 1971), p. 334.

34. Interview with Jesús Escandell, CTC foreign relations secretary, January 8, 1975, in Havana.

35. Ibid.

36. *Revolución*, September 26, p. 4, and September 28, 1962, p. 2. For the number of day-care centers, see Domínguez, *Cuba: Order and Revolution*, p. 269.

37. *Revolución,* September 28, 1962, p. 2.

38. Ibid., September 23, 1963, p. 2.

39. *Censos de población, viviendas y electoral,* pp. 204–205. The census registered 19,585 women in food and tobacco industries, 17,321 in the latter. The number of women executives, administrators, and owners of commercial enterprises was 3,294. There were 650 nurses and other health professionals, not including doctors and dentists. In 1953, 11,799 women worked in agriculture.

40. Susan Kaufman Purcell, "Modernizing Women for a Modern Society: The Cuban Case," in Ann Pescatello, ed., *Female and Male in Latin America* (Pittsburgh: University of Pittsburgh Press, 1973), p. 266, for number of women workers. I computed the labor force share from Mesa-Lago, *The Labor Force,* p. 40.

41. *Revolución,* September 18, 1963, p. 5.

42. Bertram Silverman, "The Great Debate in Retrospective: Economic Rationality and the Ethics of Revolution," in Silverman, ed., *Man and Socialism in Cuba,* p. 18.

43. Ernesto Ché Guevara, "El cuadro, columna vertebral de la revolución, septiembre de 1962," *Escritos y discursos* 6:244.

44. Fidel Castro, "Discurso en el acto de presentación del Comité Central del Partido Comunista de Cuba," *Cuba Socialista* 5 (November 1965): 61–82.

45. *El Mundo,* November 15, p. 12, and November 22, 1964, p. 11.

46. Castro, "Discurso en el acto de presentación del Comité Central," p. 81.

47. Ibid., pp. 75–76.

48. Fidel Castro, "Conclusiones del compañero Fidel Castro sobre el Poder Local," *Cuba Socialista* 5 (November 1965): 13–42.

49. Osvaldo Dorticós Torrado, "Avances institucionales de la Revolución," *Cuba Socialista* 6 (January 1966): 5.

50. Castro, "Discurso en el acto de presentación del Comité Central," pp. 79–82.

51. Fidel Castro, "Criterios de nuestra revolución," *Cuba Socialista* 5 (September 1965): 2–32.

52. Fidel Castro, "La esencia de esta hora es la técnica y el trabajo," *Cuba Socialista* 6 (October 1966): 15.

53. "La lucha contra el burocratismo," in Francisco Fernández-Santos and José Martinez, eds., *Cuba: una revolución en marcha* (Paris: Ruedo Ibérico, 1967), pp. 173–174.

54. Mesa-Lago, *The Labor Force,* p. 58.

55. "La lucha contra el burocratismo," pp. 168–187.

56. Dorticós Torrado, p. 21.

57. Ibid., pp. 2–23; p. 17 for the number of excess personnel.

58. *Granma,* July 27, 1968, p. 4.

59. Humberto Pérez, "Discurso pronunciado por Humberto Pérez en el acto de clausura del Congreso Constituyente de la Asociación Nacional de Economistas de Cuba," *Memorias: Congreso Constituyente ANEC* (Havana, 1979), pp. 119–146; Carlos Rafael Rodríguez, "Sobre la contribución del Ché al desarrollo de la economía cubana," *Cuba Socialista* 33 (May–June 1988): 1–29.

60. *Granma,* supplement, February 21, 1967, p. 3.

61. Castro, "La esencia de esta hora es la técnica y el trabajo," p. 27.

62. *Granma,* August 30, 1966, p. 4.

63. Ibid., August 26, 1966, p. 4.
64. Ibid.; and Tellería, p. 522.
65. *Granma,* August 26, 1966, p. 3.
66. Ibid.
67. Ibid., August 30, 1966, pp. 5–6.
68. Ibid., June 6, 1968, p. 2.
69. *Granma Weekly Review,* August 17, 1969, p. 1.
70. *Granma,* March 7, 1969, p. 2.
71. Ibid., September 29, 1966, p. 4.
72. *Granma Weekly Review,* October 5, 1969, p. 6.
73. Zeitlin, p. xxx.
74. Domínguez, *Cuba: Order and Revolution,* p. 268.
75. Kaufman Purcell, pp. 264–265.
76. Domínguez, *Cuba: Order and Revolution,* p. 269.
77. *Censo de población y viviendas,* pp. 679, 683.
78. Ibid., pp. 683–684. The distribution was as follows: 202,279 (41.9 percent) in social services; 101,851 (21.1 percent) in industry; 110,474 (22.9 percent) in commerce; and 39,151 (8.1 percent) in agriculture.
79. *Censo de población y viviendas,* pp. 680, 684.
80. Benigno E. Aguirre, "Women in the Cuban Bureaucracies: 1968–1974," *Journal of Comparative Family Studies* 7 (Spring 1976): 34.
81. Virginia Olesen, "Context and Posture: Notes on Socio-Cultural Aspects of Women's Roles and Family Policy in Contemporary Cuba," *Journal of Marriage and the Family* 3 (August 1971): 548–560.
82. Martin Kenner and James Petras, eds., *Fidel Castro Speaks* (New York: Grove Press, 1969), p. 248.
83. Domínguez, *Cuba: Order and Revolution,* pp. 320–321.
84. *Granma,* January 28, 1968, p. 2.
85. Héctor Ayala Castro, "Transformación de la propiedad en período 1964–1980," *Economía y desarrollo* 68 (May–June 1982): 17–20.
86. Jacques Lévesque, *The USSR and the Cuban Revolution: Soviet Ideological and Strategical Perspectives, 1959–1977* (New York: Praeger, 1978); W. Raymond Duncan, *The Soviet Union and Cuba: Interests and Influences* (New York: Praeger, 1985).
87. Fidel Castro, *Discursos,* vol. 1 (Havana: Editorial de Ciencias Sociales, 1976), p. 63.
88. *Granma,* April 1, 1969, pp. 4–5.
89. Mesa-Lago, *The Economy of Socialist Cuba,* p. 44.
90. Osvaldo Dorticós, "Discurso del Presidente de la República, Dr. Osvaldo Dorticós, en la escuela de mando del Ministerio de la Industria Ligera," *Pensamiento Crítico* 45 (October 1970): 138–156.
91. Interview with Felino Quesada, a JUCEPLAN official, on August 18, 1979 in Washington, DC; Wassily Leontieff, "Notes on a Visit to Cuba," *New York Review of Books* 21 (August 21, 1969): 15–20.
92. Fidel Castro, "Discurso pronunciado el 26 de julio de 1970—XVII aniversario del asalto al Cuartel Moncada," *Pensamiento Crítico* 45 (October 1970): 20.
93. Domínguez, *Cuba: Order and Revolution,* p. 316.
94. Castro, "Discurso pronunciado el 26 de julio de 1970," pp. 6–52.

95. Domínguez, *Cuba: Order and Revolution*, p. 276.

96. Fidel Castro, "Discurso pronunciado en la plenaria provincial de la CTC," *Pensamiento Crítico* 45 (October 1970): 108.

97. Fidel Castro, "Discurso del Comandante Fidel Castro el 23 de agosto de 1970—X aniversario de la Federación de Mujeres Cubana," *Pensamiento Crítico* 45 (October 1970): 72–74.

98. Castro, "Discurso pronunciado en la plenaria provincial de la CTC," p. 106.

99. Leo Huberman and Paul M. Sweezy, *Socialism in Cuba* (New York: Monthly Review Press, 1969); and Paul Sweezy and Charles Bettelheim, *On the Transition to Socialism* (New York: Monthly Review Press, 1971).

100. Castro, "Discurso pronunciado en la plenaria provincial de la CTC," p. 102.

101. On June 6, 1984, when the editors of *Areíto* magazine met with Carlos Rafael Rodríguez and I asked him a question about the 1960s, he used the phrase to characterize the attitude of the Cuban people at the time.

Chapter 6

1. Frank T. Fitzgerald, "A Critique of the 'Sovietization of Cuba' Thesis," *Science and Society* 42 (Spring 1978): 1–32.

2. Domínguez, *Cuba: Order and Revolution* pp. 341–378.

3. Andrew Zimbalist, "Cuban Economic Planning: Organization and Performance," in Sandor Halebsky and John M. Kirk, eds., *Cuba: Twenty-Five Years of Revolution* (New York: Praeger, 1985), p. 223.

4. *Bohemia*, July 13, 1984, p. 58. Value expanded from 500 million pesos to 1,500 million.

5. William M. LeoGrande, "Continuity and Change in the Cuban Political Elite," *Cuban Studies/Estudios Cubanos* 8 (July 1978): 1–31; Lourdes Casal and Marifeli Pérez-Stable, "Party and State in Post-1970 Cuba," in Leslie Holmes, ed., *The Withering Away of the State? Party and State Under Communism* (London: Sage, 1981), pp. 81–103; Jorge I. Domínguez, "Revolutionary Politics: The New Demands for Orderliness," in Domínguez, ed., *Cuba: Internal and International Affairs* (Beverly Hills: Sage, 1982), pp. 19–70; Archibald R. M. Ritter, "The Organs of People's Power and the Communist Party: The Nature of Cuban Democracy," in Sandor Halebsky and John M. Kirk, eds., *Cuba: Twenty-Five Years of Revolution* (New York: Praeger, 1985), pp. 270–290.

6. *Granma*, June 30, 1978, p. 4.

7. Carollee Bengelsdorf, *Between Vision and Reality: The Problem of Democracy in Cuba* (New York: Oxford University Press, 1994).

8. *Granma Weekly Review*, July 13, 1980, p. 2.

9. Bengelsdorf.

10. Bengelsdorf.

11. *Granma Weekly Review*, July 13, 1980, p. 2.

12. *Granma*, July 5, 1979, p. 1.

13. Fidel Castro, "Report of the Central Committee of the Communist Party of Cuba to the First Congress Given by Comrade Fidel Castro Ruz, First Secretary of the CC CP Cuba," in *First Congress of the Communist Party of Cuba* (Moscow: Progress Publishers, 1976), p. 133.

14. Zimbalist, "Cuban Economic Planning," p. 219.

15. Castro, "Report of the Central Committee," p. 136.

16. *Bohemia,* November 18, 1983, p. 29.

17. *Plenaria Nacional de Chequeo sobre el Sistema de Dirección y Planificación de la Economía* (Havana: JUCEPLAN, 1979); *Segunda Plenaria Nacional de Chequeo de la Implantación del SDPE* (Havana: JUCEPLAN, 1980); *Tercera Plenaria Nacional de Chequeo de la Implantación del SDPE* (Havana: Ediciones JUCEPLAN, 1982); *Dictámenes en la IV Plenaria Nacional de Chequeo del Sistema de Dirección y Planificación de la Economía* (Havana: Imprenta JUCEPLAN, 1985); Felino Quesada, "La autonomía de la empresa en Cuba y la implantación del Sistema de Dirección de la Economía," *Cuestiones de la Economía Planificada* 1 (1980): 91–99.

18. *Granma,* August 1, 1970, pp. 5–6.

19. *Granma Weekly Review,* October 24, 1971, p. 4.

20. *Trabajadores,* February 27, 1984, p. 13.

21. *Memorias XIV Congreso de la CTC* (Havana: Editorial Orbe, 1980), p. 87.

22. *XV Congreso de la CTC Memorias* (Havana: Editorial de Ciencias Sociales, 1984), p. 114.

23. Ibid., p. 82.

24. In 1985, 46.8 percent of the labor force was 34 years old or younger. Computed from *Anuario Estadístico de Cuba, 1986,* p. 201.

25. Fidel Castro and Raúl Castro, *Selecciones de discursos acerca del partido* (Havana: Editorial de Ciencias Sociales, 1975), p. 59.

26. *Granma Weekly Review,* September 26, 1972, p. 5.

27. *Memorias: Congreso de la CTC* (Havana: Central de Trabajadores de Cuba, 1974), p. 58.

28. Fidel Castro, "Discurso pronunciado en la plenaria provincial de la CTC," p. 91.

29. In 1975, I interviewed 57 workers in 15 enterprises throughout Cuba. The enterprises were among the most modern in the Cuban economy. The workers were loosely representative of vanguard workers. They averaged 10 years of education, 79 percent were vanguard workers, 27 percent of the union leaders and 22 percent of the rank and file were party members. In 1975, the labor force averaged 6 years of education, 38 percent qualified as vanguard workers, 13 percent of all union leaders and about 5 percent of the rank and file belonged to the party.

30. *Juventud Rebelde,* December 21, 1980, p. 6; and *Granma,* February 8, 1986. Supplement.

31. See Andrew Zimbalist, "On the Role of Management in Socialist Development," *World Development* 9–10 (1981): 971–977; and Linda Fuller, *Work and Democracy in Socialist Cuba* (Philadelphia: Temple University Press, 1992).

32. *Memorias XIV Congreso,* pp. 42, 44.

33. *Memorias: Congreso de la CTC,* pp. 149, 165–166.

34. *Memorias XIV Congreso,* p. 97.

35. Joaquín Benavides Rodríguez, "La ley de la distribución con arreglo al trabajo y la reforma de salarios en Cuba," *Cuba Socialista* 2 (March 1982): 62–93.

36. Zimbalist, "Incentives and Planning in Cuba," pp. 81, 84.

37. *Granma,* December 14, 1981, pp. 2–3.

38. Ibid., December 24, 1981, p. 1.

39. *XV Congreso de la CTC Memorias,* p. 25.

40. *Bohemia,* January 27, 1984, p. 49.

41. *Memorias XIV Congreso,* p. 16.

42. Ibid., p. 66.

43. *Bohemia,* November 13, 1981, p. 55.

44. *Trabajadores,* February 27, 1984, p. 7.

45. *Memorias XIV Congreso,* p. 53.

46. *Memorias: Congreso de la CTC,* p. 69.

47. *Memorias: XIV Congreso,* p. 53.

48. Roberto Veiga, "Clausura del octavo curso para directores de empresas, celebrado en la Escuela Nacional de Dirección de la Economía," *Cuestiones de la Economía Planificada* 2 (1980): 29.

49. *Memorias XIV Congreso,* p. 102.

50. *Plenaria Nacional de Chequeo sobre el Sistema de Dirección y Planificación del Economía,* pp. 36–37.

51. *Segunda Plenaria Nacional de Chequeo de la Implantación del SDPE,* p. 26.

52. Fidel Castro, *Informe Central: Tercer Congreso del Partido Comunista de Cuba* (Havana: Editora Política, 1986), p. 41.

53. Veiga, pp. 27–28.

54. *XV Congreso de la CTC Memorias,* p. 25.

55. CTC, *Dirección de Justicia Laboral* (Havana, 1981).

56. Fidel Castro, *Discurso pronunciado en la clausura del II período de sesiones de 1979 de la Asamblea Nacional del Poder Popular,* p. 6.

57. *Granma,* December 24, 1983, p. 2.

58. *Bohemia,* February 17, 1984, p. 44.

59. *XV Congreso de la CTC Memorias,* p. 35.

60. Fidel Castro, "Discurso del comandante en jefe Fidel Castro en al acto de clausura," *Memoria: II Congreso Nacional de la Federación de Mujeres Cubanas* (Havana: Editorial Orbe, 1975), p. 285.

61. Primer Congreso del Partido Comunista de Cuba, *Tesis y resoluciones* (Havana: Departamento de Orientación Revolucionaria, 1976), p. 564.

62. Lisandro Pérez, "The Family in Cuba," in Man Singh Das and Clinton J. Jesser, eds., *The Family in Latin America* (Bombay: Vikas Publishing House, 1980), pp. 235–269.

63. Isabel Larguía and John Dumoulin, "La mujer en el desarrollo: estrategias y experiencias de la revolución cubana" (Paper presented at X Latin American Congress of Sociology, Managua, 1983).

64. *Anuario Estadístico de Cuba, 1986,* pp. 519–520.

65. *Bohemia,* March 2, 1984, p. 53; *Granma Weekly Review,* March 24, 1985, p. 4.

66. *Primer Congreso del Partido Comunista de Cuba,* pp. 583–584.

67. Maritza García Alonso, "Presupuesto de tiempo de la mujer cubana: un estudio nacional, abril de 1975," *Demanda* 1 (1978): 33–58; Ana María Radaelli, "For the Full Equality of Women," *Cuba International* (July 1985): 13–17; Vilma Espín, "La batalla por el ejercicio pleno de la igualdad de la mujer: acción de los comunistas," *Cuba Socialista* 6 (1986): 27–68.

68. Castro, *Informe Central,* p. 78.

69. *Memoria: II Congreso Nacional de la Federación de Mujeres Cubanas* (Havana: Editorial Orbe, 1975), pp. 173–174.

70. *Granma,* June 1, 1976, p. 4.

71. On January 12, 1977, I asked Vilma Espín about the 1976 resolution when she met with the Antonio Maceo Brigade.

72. Vilma Espín, "Central Report Rendered by Comrade Vilma Espín, President of the Federation of Cuban Women at the Federation's Third Congress," *Boletín FMC* (Havana, 1980), pp. 20–21.

73. Carollee Bengelsdorf, "On the Problem of Studying Cuban Women," in Andrew Zimbalist, ed., *Cuban Political Economy: Essays in Cubanology* (Boulder: Westview Press, 1988), p. 128.

74. Espín, p. 39.

75. Fidel Castro, "Speech Delivered by Commander in Chief Fidel Castro at the Closing Session of the Federation's Third Congress," *Boletín FMC* (Havana, 1980), p. 36.

76. Sergio Díaz-Briquets, "Age-Structure, Fertility Swings, and Socio-economic Development in Cuba," in Roca, ed., *Socialist Cuba,* pp. 159–174.

77. Muriel Nazzari, "The 'Woman Question' in Cuba: An Analysis of Material Constraints on its Solution," *Signs* 2 (1983): 246–263.

78. Radaelli, p. 16.

79. *Censo de población y viviendas de 1981,* vol. 16, pt. 2, pp. 299, 303.

80. *Trabajadores,* February 27, 1984, pp. 8–9.

81. Computed from *Anuario Estadístico de Cuba, 1986,* p. 203.

82. Computed from ibid., pp. 196, 200.

83. *Bohemia,* November 16, 1984, p. 40.

84. Espín, "La batalla por el ejercicio pleno de la igualdad de la mujer," pp. 59–62.

85. *Anuario Estadístico de Cuba, 1986,* p. 513.

86. Castro, "Report of the Central Committee," p. 232.

87. Fidel Castro, *Main Report: Second Congress of the Communist Party of Cuba* (New York: Center for Cuban Studies, 1981), p. 27.

88. Castro, *Informe Central,* p. 99.

89. Jorge I. Domínguez, "Blaming Itself, Not Himself: Cuba's Political Regime After the Third Party Congress," in Roca, ed., *Socialist Cuba: Past Interpretations and Future Challenges,* pp. 3–10; Supplement of *Granma,* February 8, 1986, pp. 1–6.

90. Massimo Cavallini, "La revolución es una obra de arte que debe perfeccionarse," *Pensamiento Propio* 33 (May–June, 1986): 43.

91. Castro, "Report of the Central Committee," p. 239.

92. Domínguez, "Blaming Itself, Not Himself," p. 7.

93. Castro, "Report of the Central Committee," p. 232.

94. Castro, *Informe Central,* pp. 40–41.

Chapter 7

1. On the anniversary of Playa Girón, Fidel Castro berated the new wave of mercenaries that threatened *la patria* and called for a renewed struggle against the would-be capitalists. After April, the rectification began to take shape in various forums. At a gathering of cooperative farmers, Castro announced the

banning of the free markets (*Granma Weekly Review,* June 1, 1986, pp. 3–4). Meeting with enterprise representatives from the two Havana provinces, he assailed the "liberal-bourgeois" period of "bitter experiences" and called for a "strategic counteroffensive" (*Granma,* June 27, 1986, pp. 1–3). Throughout July and August, enterprise meetings in the other provinces continued elaborating on the problems of the SDPE (*Granma,* July 8, p. 11; July 9, pp. 1, 3; July 10, p. 3; August 11, p. 3; and August 19, 1986, p. 3). On July 26, Castro railed against the "generalized stupidities" that money was the principal motivation of people and exalted the value of volunteer work (*Granma Weekly Review,* August 2, 1986, p. 3). In December, the party congress closed with a call to renew moral principles and regenerate *conciencia* (*Granma,* December 1, 2, 1986; and Supplement, December 5, 1986).

2. *Plan de acción contra las irregularidades administrativas y los errores y debilidades del Sistema de Dirección de la Economía,* July 17, 1986.

3. *Bohemia,* December 14, 1984, pp. 51–63.

4. *Granma Weekly Review,* February 15, 1987, p. 5.

5. *Granma,* July 5, p. 2; August 3, p. 5; November 3, 1986, p. 2; and December 13, 1987, p. 8; *Granma Weekly Review,* February 15, p. 5; and May 17, 1987, p. 9.

6. *Granma,* December 31, 1986, pp. 1–2.

7. *Granma Weekly Review,* June 21, 1987, p. 4.

8. Pedro Monreal, "Cuba y la nueva economía mundial: el reto de la inserción en América Latina y el Caribe," *Cuadernos de Nuestra América* 8 (January–June 1991): 64.

9. *Cuba en el mes,* August 1990, p. 27.

10. *Granma,* December 26, 1986, p. 5.

11. Computed from *Anuario Estadístico de Cuba, 1986,* p. 196.

12. *Granma Weekly Review,* July 5, 1987, p. 5; Fidel Castro, *Discurso pronunciado en la clausura del III congreso de los Comités de Defensa de la Revolución* (September 28, 1986), p. 6; *Granma,* December 1, 1986, p. 4.

13. *Bohemia,* March 28, 1986, pp. 26–36; and *Granma,* July 5, 1986, p. 1.

14. *Granma,* July 22, 1986, p. 1.

15. *Granma Weekly Review,* June 18, 1986, p. 3.

16. *Granma,* July 11, 1988, p. 4; and *Granma Weekly Review,* August 7, 1988, p. 3.

17. *Granma,* July 4, 1986, p. 3.

18. *Granma Weekly Review,* January 29, 1989, p. 5.

19. *Anuario Estadístico de Cuba 1988,* p. 282. Housing construction declined to 35,659 in 1988, with private construction representing about 20 percent of the total.

20. *Granma Weekly Review,* January 22, 1989, Supplement.

21. Ibid., February 7, 1988, p. 9.

22. *Bohemia,* October 30, 1987, p. 20.

23. *Granma,* December 10, 1988, p. 2.

24. Ibid., December 27, 1986, p. 4.

25. Ibid., December 10, 1988, p. 2.

26. *Granma Weekly Review,* January 8, 1989, p. 9.

27. *Granma,* December 1, p. 2; and December 5, 1986, Supplement; Janu-

ary 24, p. 3; and March 21, 1989, p. 3; Fidel Castro, "En la reunión informativa del Comité Provincial del partido de Ciudad de La Habana," *Cuba Socialista* 26 (March–April 1987): 1–42, and "En la asamblea provincial del partido de Ciudad de La Habana," *Cuba Socialista* 31 (January–February 1988): 1–37.

28. Castro, *Informe Central*, p. 99.

29. *Granma*, December 25, p. 2; and December 26, 1986, pp. 2–3.

30. *Granma Weekly Review*, December 13, 1987, p. 9.

31. *Granma*, December 2, 1986; *Granma Weekly Review*, February 1, 1987, pp. 2–4.

32. *Granma Weekly Review*, February 1, 1987, pp. 2–4.

33. *Granma*, December 25, 1988, p. 9.

34. Castro, "En la reunión informativa del Comité Provincial del partido de Ciudad de La Habana," p. 13.

35. The Ochoa-de la Guardia public proceedings, their press coverage, and other related documents can be found in *Causa 1/89: Fin de la conexión cubana* (Havana: Editorial José Martí, 1989). Andrés Oppenheimer, *Castro's Final Hour: The Secret Story Behind the Coming Downfall of Communist Cuba* (New York: Simon & Schuster, 1992), pp. 17–163, is an insightful analysis of the drug trafficking trials of 1989.

36. *Granma*, June 16, p. 1; June 29, p. 1; July 14, p. 1; and July 31, 1989, p. 11; and *Granma Weekly Review*, August 20, 1989, p. 3.

37. *Granma*, August 30, 1989, pp. 3–5. In January 1991, Abrantes suffered a fatal heart attack while under confinement.

38. José Luis Llovió-Menéndez, *Insider: My Hidden Life as a Revolutionary in Cuba* (Toronto and New York: Bantam Books, 1988).

39. *Granma Weekly Review*, September 10, 1989, pp. 1, 11.

40. *Granma*, February 17, 1990, p. 1; *Granma Weekly Review*, March 25, 1990, pp. 2–3.

41. *Granma Weekly Review*, March 26, 1989, p. 9.

42. *XV Congreso de la CTC*, p. 194.

43. The January 1990 CTC congress published two documents: *Los sindicatos en el proceso de rectificación: documento base* (n.d.), and *Los sindicatos en el proceso de rectificación: informe central* (n.d.).

44. *Proyecto de tesis V congreso de la Federación de Mujeres Cubanas* (n.d.). Women accounted for 51.4 percent of local trade union leaders and 38.2 percent of the labor force. See pp. 10–11 of the *documento base* and pp. 15–16 of the *informe central* to the CTC congress of 1990 for the references to working women without mentioning the FMC.

45. See *Granma*, June 16, p. 7; June 19, p. 3; June 20, p. 3; and June 26, 1990, p. 3; and *Granma Weekly Review*, June 18, 1990, p. 1. The standard criticism pointed out the overlapping functions between the FMC and the Committees for the Defense of the Revolution. CDR President Sixto Batista was also dropped from the Politburo in the October PCC congress.

46. For the FMC congress, see *Granma*, March 6, pp. 1, 3–4; March 7, pp. 1–2, 3–5; and March 8, 1990, pp. 1, 3; and *Granma Weekly Review*, March 18, 1990, pp. 7–12.

47. *Granma*, April 13, 1990, p. 1.

48. Ibid., June 23, 1990, pp. 4–5.

49. *Granma Weekly Review*, March 25, p. 3; April 22, 1990, p. 9.

50. Personal communication to the author by several colleagues who are PCC members during a visit to Havana in May 1991.

51. *Granma,* January 6, 1990, p. 1.

52. Ibid., January 28, 1991, p. 3.

53. *Granma Weekly Review,* October 14, 1990, p. 9.

54. *Cuba en el mes,* July 1991, p. 8.

55. *Granma,* January 28, 1991, p. 3.

56. *Cuba en el mes,* August 1991, p. 65.

57. *Granma,* October 13, 1991, p. 7.

58. Foreign Broadcast Information Service, *Latin America—Cuba: Fourth Congress of the Cuban Communist Party,* October 15, 1991, pp. 1–24.

59. Ibid., pp. 3–24 for Castro speech; pp. 9–15 for the list of shortfalls in Soviet delivery.

60. *Granma,* October 16, 1991, p. 3.

61. For the resolutions, see PCC statutes, *Granma,* October 13, 1991, p. 7; PCC program, *Granma,* October 14, 1991, p. 6; Popular Power and foreign policy, *Granma,* October 16, 1991, pp. 3, 6; Central Committee special powers, *Bohemia,* October 18, 1991, pp. 32–33; economic development, Pablo Alfonso, *Los fieles de Castro* (Miami: Ediciones Cambio, 1991), pp. 212–223.

62. Foreign Broadcast Information Service, pp. 40–41; and computed from Alfonso, pp. 23–162.

63. Foreign Broadcast Information Service, p. 25.

64. Although the human rights movement was small and had little domestic impact, the signs of popular discontent were telling and extensive: signs of "¡Abajo Fidel!" painted on walls; clashes between youths and the police; dock workers refusing to load sacks of rice for export because of domestic scarcities; the film institute challenging a party directive to merge with state television; intellectuals signing an open letter to the leadership demanding reform; purges at the University of Havana and other higher education institutions; citizens raiding planted fields; near-riots in front of the special hard-currency stores; workers refusing to join the rapid response brigades the government created to quell dissent. See *Miami Herald,* January 12, 1990, p. 10; *El Nuevo Herald,* May 29, p. 1, December 16, 1990, p. 3, January 5, p. 3, February 10, p. 3, March 13, p. 3, June 24, p. 1, and June 30, 1991, p. 1; *New York Times,* June 29, 1991, p. 2; Liz Balmaseda, "Castro's Convertibles," *Tropic,* April 14, 1991, pp. 9–15, 18–19; and Oppenheimer, pp. 267–282, 304–337, 401–423.

Conclusion

1. Quoted in Thomas, p. 101.

2. For the constitutional reforms, see Hugo Azcuy, Rafael Hernández, and Nelson P. Valdés, "Reforma constitucional cubana," *Cuba en el mes* (July 1992).

3. In March, Carlos Aldana expressed this view and even named two of the dissidents—Elizardo Sánchez and Gustavo Arcos. Foreign Broadcast Information Service, "Latin America 1992" (March 11, 1992), pp. 5–6. In September, Aldana lost his position in the Politburo and his party membership. The official explanation was that he had used bad judgment in the conduct of official business transactions.

4. Foreign Broadcast Information Service, (September 14, 1992), pp. 7–8;

Juventud Rebelde, October 18, 1992, p. 2; *Trabajadores*, October 19, 1992, p. 2.

5. CUBAFAX, December 31, 1992 for the unofficial estimates; Spanish News Agency, EFE, February 24, 1993 for the official figure. Immediately following the December elections, dissidents in Cuba and the international press reported that about a third of the electorate cast invalid ballots. Initially, government spokesmen did not deny that figure nor did they offer an official figure. Two months later, then-National Assembly President Juan Escalona cited the less than 15 percent figure.

6. *El Nuevo Herald*, March 2, 1993, p. 3.

7. Spanish News Agency EFE, February 25, 1993.

8. In February 1993, however, President Castro told Diane Sawyer on ABC's *Prime Time Live* that he would decline to run for his National Assembly seat when the current term expires in 1998 if the United States changed its policy toward Cuba.

Select Bibliography

Agrupación Católica Universitaria. *¿Por qué reforma agraria?* Havana, 1957.

Aguirre, Benigno. "Women in the Cuban Bureaucracies: 1968–1974." *Journal of Comparative Family Studies* 7 (Spring 1976): 23–40.

Alfonso, Pablo. *Los fieles de Castro.* Miami: Ediciones Cambio, 1991.

Alienes Urosa, Julián. *Características fundamentales de la economía cubana.* Havana: Banco Nacional, 1950.

América en Cifras. 8 vols. Washington, DC: Panamerican Union, 1960.

Anderson, Benedict. *Imagined Communities: Reflections on the Origins and Spread of Nationalism.* London: Verso Editions, 1983.

Aya, Rod. *Rethinking Revolutions and Collective Violence: Studies on Concept, Theory, and Method.* Amsterdam: Het Spinhuis, 1990.

Ayala Castro, Héctor. "Transformación de propiedad en el perído 1964–1980." *Economía y Desarrollo* 68 (May–June 1982): 11–25.

———. "Transformaciones de propiedad, control obrero e intervención de empresas en Cuba (1959–1960)." *Economía y Desarrollo* 47 (May–June 1978): 44–69.

Azicri, Max. *Cuba: Politics, Economics and Society.* London: Francis Pinter, 1988.

Barkin, David P. "Cuban Agriculture: A Strategy of Economic Development." In *Cuba: The Logic of the Revolution,* edited by David P. Barkin and Nita Rous Manitzas. Andover, MA: Warner Modular Publications, 1973.

Becker, David. *The New Bourgeoisie and the Limits of Dependency.* Princeton: Princeton University Press, 1983.

Benavides Rodríguez, Joaquín. "La ley de la distribución con arreglo al trabajo y la reforma de salarios en Cuba." *Cuba Socialista* 2 (March 1982): 62–93.

Bengelsdorf, Carollee. *Between Vision and Reality: The Problem of Democracy in Cuba*. New York: Oxford University Press, 1994.

———. "On the Problem of Studying Cuban Women." In *Cuban Political Economy: Essays in Cubanology*, edited by Andrew Zimbalist, 119–136. Boulder: Westview Press, 1988.

Benjamin, Jules Robert. *The United States and Cuba: Hegemony and Dependent Development, 1880–1934*. Pittsburgh: Univeristy of Pittsburgh Press, 1974.

Bergquist, Charles. *Labor in Latin America*. Stanford: Stanford University Press, 1986.

Blackburn, Robin. "Prologue to the Cuban Revolution." *New Left Review* 21 (October 1963): 52–91.

Blasier, Cole. "COMECON in Cuban Development." In *Cuba in the World*, edited by Cole Blasier and Carmelo Mesa-Lago, 225–255. Pittsburgh: University of Pittsburgh Press, 1979.

Bonachea, Ramón L., and Marta San Martín. *The Cuban Insurrection: 1952–1959*. New Brunswick, NJ: Transaction Books, 1974.

Bonachea, Rolando E., and Nelson P. Valdés, eds. *Revolutionary Struggle (1947–1958). Vol. 1 of Selected Works of Fidel Castro*. Cambridge: MIT Press, 1972.

———, eds., *Ché: Selected Works of Ernesto Guevara*. Cambridge: MIT Press, 1969.

Boorstein, Edward. *The Economic Transformation of Cuba*. New York: Monthly Review Press, 1968.

Boti, Regino. "El plan de desarrollo económico de 1962." *Cuba Socialista* 1 (December 1961): 19–32.

Boti, Regino, and Felipe Pazos. *Algunos aspectos del desarrollo económico de Cuba. Tésis del M-26-7*. Havana: Delegación del Gobierno en el Capitolio Nacional, 1959.

Brown, James G. *The International Sugar Industry: Developments and Prospects*. Washington, DC: World Bank, 1987.

Brundenius, Claes. "Development and Prospects of Capital Goods Production in Revolutionary Cuba." In *Cuba's Socialist Economy Toward the 1990s*, edited by Andrew Zimbalist, 97–114. Boulder: Lynne Rienner 1987.

———. *Revolutionary Cuba: The Challenge of Economic Growth with Equity*. Boulder: Westview Press, 1984.

Brundenius, Claes, and Andrew Zimbalist. "Cuban Growth: A Final Word." *Comparative Economic Studies* 27 (Winter 1985): 83–84.

———. "Cuban Economic Growth One More Time: A Response to 'Imbroglios.'" *Comparative Economic Studies* 27 (Fall 1985): 115–131.

———. "Recent Studies on Cuban Economic Growth: A Review." *Comparative Economic Studies* 27 (Spring 1985): 22–46.

Brunner, Heinrich. *Cuban Sugar Policy from 1963 to 1970*. Pittsburgh: University of Pittsburgh Press, 1977.

Cámara de Comercio de la República de Cuba and Asociación Nacional de Industriales de Cuba, eds. *Conferencia para el progreso de la economía nacional*, Havana, 1949.

Cardoso, Fernando Henrique, and Enzo Faletto. *Dependency and Development in Latin America*. Berkeley: University of California Press, 1979.

Carrillo, Justo. *Cuba 1933: estudiantes, yanquis y soldados*. Coral Gables: University of Miami, 1985.

Carvajal, Juan F. "Observaciones sobre la clase media en Cuba." In *Materiales para el estudio de la clase media en la América Latina*, edited by Theo R. Crevenna, 30–44. Vol. 2. Washington, DC: Panamerican Union, 1950–1951.

Casal, Lourdes, and Marifeli Pérez-Stable. "Party and State in Post-1970 Cuba." In *The Withering Away of the State? Party and State under Communism*, edited by Leslie Holmes, 81–103. London: Sage, 1981.

Castro, Fidel. "En la asamblea provincial del partido de Ciudad de La Habana." *Cuba Socialista* 31 (January–February 1988): 1–37.

———. "En la reunión informativa del Comité Provincial del partido de Ciudad de La Habana." *Cuba Socialista* 26 (March–April, 1987): 1–42.

———. *Informe Central: Tercer Congreso del Partido Comunista de Cuba*. Havana: Editora Política, 1986.

———. *Discurso pronunciado en la clausura del III congreso de los Comités de Defensa de la Revolución*. September 28, 1986.

———. *Main Report: Second Congress of the Communist Party of Cuba*. New York: Center for Cuban Studies, 1981.

———. "Speech Delivered by Commander in Chief Fidel Castro at the Closing Session of the Federation's Third Congress." *Boletín FMC* (1980), pp. 31–45.

———. *Discurso pronunciado en la clausura del II período de sesiones de 1979 de la Asamblea Nacional del Poder Popular*. December 27, 1979.

———. *Discursos*. 2 vols. Havana: Editorial de Ciencias Sociales, 1976.

———. "Report of the Central Committee of the Communist Party of Cuba to the First Congress Given by Comrade Fidel Castro Ruz, First Secretary of the CC CP Cuba." In *First Congress of the Communist Party of Cuba*. Moscow: Progress Publishers, 1976: 16–279.

———. "Discurso del Comandante Fidel Castro el 23 de agosto de 1970—X aniversario de la Federación de Mujeres Cubanas." *Pensamiento Crítico* 45 (October 1970): 53–85.

———. "Discurso pronunciado el 26 de julio de 1970—XVII aniversario del asalto al Cuartel Moncada." *Pensamiento Crítico* 45 (October 1970): 6–52.

———. "Discurso pronunciado en la plenaria provincial de la CTC." *Pensamiento Crítico* 45 (October 1970): 86–115.

———. "La esencia de esta hora es la técnica y el trabajo." *Cuba Socialista* 6 (October 1966): 4–30.

———. "Conclusiones del compañero Fidel Castro sobre el Poder Local." *Cuba Socialista* 5 (November 1965): 13–42.

———. "Discurso de Fidel Castro en la presentación del Comité Central del Partido Comunista de Cuba." *Cuba Socialista* 5 (November 1965): 61–82.

———. "Criterios de nuestra revolución." *Cuba Socialista* 5 (September 1965): 2–32.

———. "Discurso del Primer Ministro en el Comité Provincial de Matanzas." *Cuba Socialista* 2 (May 1962): 1–27.

Castro, Fidel, and Raúl Castro. *Selecciones de discursos acerca del partido*. Havana: Editorial de Ciencias Sociales, 1975.

Causa 1/89: Fin de la conexión cubana. Havana: Editorial José Martí, 1989.

Cavallini, Massimo. "La revolución es una obra de arte que debe perfeccionarse." *Pensamiento Propio* 33 (May–June 1986): 39–43.

Cepero Bonilla, Raúl. *Escritos económicos.* Havana: Editorial de Ciencias Sociales, 1983.

———. *Azúcar y abolición.* Havana: Editorial de Ciencias Sociales, 1971.

———. *Política azucarera (1952–1958).* Mexico: Editora Futuro, 1958.

Collier, Ruth Berins, and David Collier. *Shaping the Political Arena: Critical Junctures, the Labor Movement, and Regime Dynamics in Latin America.* Princeton: Princeton University Press, 1991.

Consejo Nacional de Economía. *El programa económico de Cuba.* Havana, 1955.

Cuban Economic Research Project. *Labor Conditions in Communist Cuba.* Miami: University of Miami Press, 1963.

Cuban Women, 1975–1979. Havana, 1980.

de Armas, Ramón. *La revolución pospuesta,* Havana: Editorial de Ciencias Sociales, 1975.

de Janvry, Alain. *The Agrarian Question and Reformism in Latin America.* Baltimore: Johns Hopkins University Press, 1981.

de la Cuesta, Leonel-Antonio, ed. *Constituciones cubanas: desde 1812 hasta nuestros días.* New York: Ediciones Exilio, 1974.

del Aguila, Juan. *Cuba: Dilemmas of a Revolution.* Boulder: Westview Press, 1984.

Di Tella, Torcuato S. *Latin American Politics: A Theoretical Framework.* Austin: University of Texas Press, 1990.

Díaz-Briquets, Sergio. "Age-Structure, Fertility Swings, and Socioeconomic Development." In *Socialist Cuba: Past Interpretations and Future Challenges,* edited by Sergio Roca, 159–174. Boulder: Westview Press, 1988.

———. "Regional Differences in Development and Living Standards in Revolutionary Cuba." *Cuban Studies/Estudios Cubanos* 18 (1988): 45–63.

———. *The Health Revolution in Cuba.* Austin: University of Texas Press, 1983.

Dictámenes en la IV Plenaria Nacional de Chequeo del Sistema de Dirección y Planificación de la Economía. Havana: Imprenta JUCEPLAN, 1985.

Domínguez, Jorge I. *To Make the World Safe for Revolution: Cuba's Foreign Policy.* Cambridge: Harvard University Press, 1989.

———. "Blaming Itself, Not Himself: Cuba's Political Regime After the Third Party Congress." in *Socialist Cuba: Past Interpretations and Future Challenges,* edited by Sergio Roca, 3–10. Boulder: Westview Press, 1988.

———. "Seeking Permission to Build a Nation: Cuban Nationalism and U.S. Response Under the First Machado Presidency." *Cuban Studies/Estudios Cubanos* 16 (1986): 33–48.

———. "Revolutionary Politics: The New Demand for Orderliness." In *Cuba: Internal and International Affairs,* edited by Jorge I. Domínguez, 19–70. Beverly Hills: Sage, 1982.

———. *Cuba: Order and Revolution.* Cambridge: Harvard University Press. 1978.

Dorticós Torrado, Osvaldo. "Discurso del Presidente de la República, Dr. Osvaldo Dorticós, en la escuela de mando del Ministerio de la Industria Ligera." *Pensamiento Crítico* 45 (October 1970): 138–156.

———. "Avances institucionales de la Revolución." *Cuba Socialista* 6 (January 1966): 2–23.

Draper, Theodore. *Castroism: Theory and Practice.* New York: Praeger, 1965.

————. *Castro's Revolution: Myths and Realities.* New York: Praeger, 1962.

Duncan, W. Raymond. *The Soviet Union and Cuba: Interests and Influences.* New York: Praeger, 1985.

Eckstein, Susan. "State and Market Dynamics in Castro's Cuba." In *States versus Markets in the World System,* edited by Peter Evans, Dietrich Rueschmeyer, and Evelyne Huber Stephens, 217–245. Beverly Hills: Sage 1985.

Edquist, Charles. *Capitalism, Socialism and Technology.* London: Zed Books, 1985.

"El desarrollo económico de Cuba," *Revista del Banco Nacional* 2 (March 1956): 273–276.

"El desarrollo industrial de Cuba," *Cuba Socialista* 6 (April 1966): 128–183.

Erisman, H. Michael. *Cuba's International Relations: The Anatomy of a Nationalistic Foreign Policy.* Boulder: Westview Press, 1985.

Espín, Vilma. "La batalla por el ejercicio pleno de la igualdad de la mujer: acción de los comunistas." *Cuba Socialista* 6 (1986): 27–68.

————. "Central Report Rendered by Comrade Vilma Espín, President of the Federation of Cuban Women at the Federation's Third Congress." *Boletín FMC* (1980): 9–28.

Evans, Peter. *Dependent Development: The Alliance of Multinational, State, and Local Capital in Brazil.* Princeton: Princeton University Press, 1979.

Evans, Peter B., and John D. Stephens, "Development and the World Economy." In *Handbook of Sociology,* edited by Neil J. Smelser, 739–773. Newbury Park, CA: Sage 1988.

Fagen, Richard. *The Transformation of Political Culture in Cuba.* Stanford: Stanford University Press, 1969.

Farber, Samuel. *Revolution and Reaction in Cuba, 1933–1960.* Middletown, CT: Wesleyan University Press, 1976.

Feinsilver, Julie M. "The Politics of Health in Cuba in the Age of Perestroika: Capitalizing on Biotechnology and Medical Exports." *Cuban Studies/ Estudios Cubanos* 22 (1992): 79–111.

Fernández Font, Marcelo. *Cuba y la economía azucarera mundial.* Havana: Instituto Superior de Relaciones Internacionales, 1986.

Fernández Plá, Francisco. "Política social apropriada para el fomento de la producción nacional." In *Conferencia para el progreso de la economía nacional,* edited by Cámara de Comercio de la República de Cuba and, Asociación Nacional de Industriales de Cuba, 155–173. Havana, 1949.

Fernández-Santos, Francisco, and José Martínez, eds. *Cuba: una revolución en marcha.* Paris: Ruedo Ibérico, 1967.

Feuer, Carl Henry. "The Performance of the Cuban Sugar Industry, 1981–1985." In *Cuba's Socialist Economy Toward the 1990s,* edited by Andrew Zimbalist, 69–83. Boulder: Lynne Rienner, 1987.

First Congress of the Communist Party of Cuba. Moscow: Progress Publishers, 1976.

Fitzgerald, Frank T. "A Critique of the 'Sovietization of Cuba' Thesis." *Science & Society* 42 (Spring 1978): 1–32.

Font, Mauricio. *Coffee, Contention, and Change in the Making of Modern Brazil.* Oxford and Cambridge: Basil Blackwell, 1990.

Foreign Policy Association. *Problemas de la nueva Cuba.* New York: J. J. Little and Ives, 1935.

Franqui, Carlos. *Retrato de familia con Fidel*. Barcelona: Editorial Seix Barral, 1981.
———. *Diary of the Cuban Revolution*. New York: Viking Press, 1980.
Fuller, Linda. *Work and Democracy in Socialist Cuba*. Philadelphia: Temple University Press, 1992.
Gallo, Carmenza. *Taxes and State Power: Political Instability in Bolivia, 1900–1952*. Philadelphia: Temple University Press, 1991.
García Alonso, Maritza. "Presupuesto de tiempo de la mujer cubana: un estudio nacional, abril de 1975." *Demanda* 1 (1978): 33–58.
Gereffi, Gary. *The Pharmaceutical Industry and Dependency in the Third World*. Princeton: Princeton University Press, 1983.
Gilly, Adolfo. "Inside the Cuban Revolution." *Monthly Review* 16 (October 1964): 13–20.
Goldenberg, Boris. *The Cuban Revolution and Latin America*. New York: Praeger, 1965.
Goldstone, Jack A. "The Comparative and Historical Study of Revolutions." *Annual Review of Sociology* 8 (1982): 187–207.
Gómez, Isidro. "El Partido Comunista de Cuba." Paper presented at the seminar of the Institute for Cuban Studies, Washington DC. August 16–18, 1979.
González, Edward. *Cuba under Castro: The Limits of Charisma*. Boston: Houghton Mifflin, 1974.
Gramsci, Antonio. *Letters from Prison*, selected, translated, and introduced by Lynne Lawner. New York: Harper & Row, 1973.
Grupo Cubano de Investigaciones Económicas. *Un estudio sobre Cuba*. Miami: University of Miami Press, 1963.
Guerra, José Antonio, "Necesidad de organizar sobre bases permanentes y nacionales el estudio y determinación de la política comercial internacional de Cuba." In *Conferencia para el progress de la economía nacional*, edited by Cámara de Comercio de la República de Cuba and Asociación Nacional de Industriales de Cuba, 211–222. Havana, 1949.
Guerra, Ramiro. *Azúcar y población en las Antillas*. Havana: Editorial de Ciencias Sociales, 1976.
Guevara, Ernesto. "El cuadro, columna vertebral de la revolución, septiembre de 1962." In *Escritos y discursos*, 6:239–245. Havana: Editorial de Ciencias Sociales, 1977.
———. "Discurso a la clase obrera, 14 de junio de 1960." In *Escritos y discursos*, 4:127–154. Havana: Editorial de Ciencias Sociales, 1977.
———. "Discurso clausura del Consejo Nacional de la CTC, 15 de abril de 1962." In *Escritos y discursos*, 6:129–146. Havana: Editorial de Ciencias Sociales, 1977.
———. "Discurso en la Convención Nacional de los Consejos Técnicos Asesores." In *Escritos y discursos*, 5:37–40. Havana: Editorial de Ciencias Sociales, 1977.
———. "Discusión colectiva: decisión y responsabilidad unicas." In *Escritos y discursos*, 5:191–209. Havana: Editorial de Ciencias Sociales, 1977.
———. "El socialismo y el hombre en Cuba." In Escritos y discursos, 8:253–272. Havana: Editorial de Ciencias Sociales, 1977.
———. "The Meaning of Socialist Planning." In *Man and Socialism in Cuba: The Great Debate*, 98–110. New York: Atheneum Press, 1971.

————. "The Alliance for Progress." In *Ché: Selected Works of Ernesto Guevara,* edited by Rolando Bonachea and Nelson P. Valdés, 265–296. Cambridge: MIT Press, 1969.

Guiral, Enrique J. "Orientación de la legislación social para el desarrollo de la economía cubana: proyecciones de la reglamentación estatal del trabajo en una economía basada en la empresa libre." In *Conferencia para el progreso de la economía nacional,* edited by Cámara de Comercio de la República de Cuba and Asociación Nacional de Industriales de Cuba, 123–134. Havana, 1949.

Gunder Frank, André. *Capitalism and Underdevlopment in Latin America.* New York: Monthly Review Press, 1967.

Gutiérrez, Gustavo. *El desarrollo económico de Cuba.* Havana: Junta Nacional de Economía, 1952.

————. "Necesidad de adoptar una política exterior." In *El problema económico de Cuba,* edited by Gustavo Gutiérrez, 17–32. Havana: Molina, 1931.

Hamburg, Jill. "Housing Policy in Revolutionary Cuba." In *Critical Perspectives on Housing,* edited by Rachel Bratt, Chester Hartman, and Ann Meyerson, 586–624. Philadelphia: Temple University Press, 1986.

Hernández, Roberto E. and Carmelo Mesa-Lago, "Labor Organization and Wages." In *Revolutionary Change in Cuba,* edited by Carmelo Mesa-Lago, 209–249. Pittsburgh: University of Pittsburgh Press, 1971.

Hobsbawm, Eric. "Revolution." In *Revolution in History,* edited by Roy Porter and Mikulás Teich, 5–46. Cambridge: Cambridge University Press, 1986.

Huberman, Leo, and Paul Sweezy. *Socialism in Cuba.* New York: Monthly Review Press, 1969.

————. *Cuba: Anatomy of a Revolution.* New York: Monthly Review Press, 1961.

Ibarra, Jorge. *José Martí: dirigente político y revolucionario.* Havana: Editorial de Ciencias Sociales, 1980.

————. *Ideología mambisa.* Havana: Instituto del Libro, 1972.

Instituto de Historia del Movimiento Comunista y de la Revolución Socialista de Cuba. *Historia del movimiento obrero cubano.* 2 vols. Havana: Editora Política, 1985.

International Bank for Reconstruction and Development. *Report on Cuba.* Baltimore: Johns Hopkins University Press, 1951.

Jenks, Leland H. *Nuestra colonia de Cuba.* Buenos Aires: Editorial Palestra, 1959.

Kaufman Purcell, Susan. "Modernizing Women for a Modern Society: The Cuban Case." In *Female and Male in Latin America,* edited by Ann Pescatello, 257–271. Pittsburgh: University of Pittsburgh Press, 1973.

Kay, Cristóbal. *Latin American Theories of Development and Underdevelopment.* London: Routledge, 1989.

Kenner, Martin and James Petras, eds. *Fidel Castro Speaks.* New York: Grove Press, 1969.

Kimmel, Michael S. *Revolution: A Sociological Interpretation.* Philadelphia: Temple University Press, 1990.

Kirk, John M. *José Martí: Mentor of the Cuban Nation.* Tampa: University Presses of Florida, 1983.

"La reforma arancelaria." *Revista del Banco Nacional* 1 (January 1958): 5–21.

Larguía, Isabel, and John Dumoulin. "La mujer en el desarrollo: estrategias y experiencias de la revolución cubana." Paper presented at the Tenth Latin American Congress of Sociology. Managua, Nicaragua, 1983.

Leal, Juan Felipe. "Las clases sociales en Cuba en vísperas de la revolución." *Revista Mexicana de Ciencia Política* 19 (October–December 1973): 99–109.

Lenin, V. I. *Selected Works: July 1918 to March 1923*. New York: International Publishers, 1967.

LeoGrande, William M. "Cuban Dependency: A Comparison of Pre-Revolutionary and Post-Revolutionary International Relations." *Cuban Studies/Estudios Cubanos* 9 (July 1979): 1–28.

———. "Continuity and Change in the Cuban Political Elite." *Cuban Studies/Estudios Cubanos* 8 (July 1978): 1–32.

Leontieff, Wassily. "Notes on a Visit to Cuba." *New York Review of Books* 21 (August 21, 1969): 15–20.

Lévesque, Jacques. *The USSR and the Cuban Revolution: Soviet Ideological and Strategical Perspectives, 1959–1977*. New York: Praeger, 1978.

Liss, Sheldon B. *The Roots of Revolution: Radical Thought in Cuba*. Lincoln: University of Nebraska Press, 1987.

Llovió-Menéndez, José Luis. *Insider: My Hidden Life as a Revolutionary in Cuba*. Toronto and New York: Bantam Books, 1988.

López-Segrera, Francisco. *Raíces históricas de la revolución cubana (1868–1959)*. (Havana: Ediciones Unión, 1980).

———. *Cuba: capitalismo dependiente y subdesarrollo (1510–1959)*. Havana: Casa de las Américas, 1972.

Los sindicatos en el proceso de rectificación: documento base.

Los sindicatos en el proceso de rectificación: informe central.

Luzón, José L. "Housing in Socialist Cuba: An Analysis Using Cuban Censuses of Population and Housing." *Cuban Studies/Estudios Cubanos* 18 (1988): 65–83.

Machado, Luis. *La isla de corcho: ensayo de economía cubana*. Havana: Mazo, Caso, 1936.

———. "Necesidad de adoptar una política de comercio exterior." In *El problema económico de Cuba*, edited by Gustavo Gutiérrez, 35–80. Havana: Molina, 1931.

Martí, Jorge L. "Class Attitudes in Cuban Society on the Eve of the Revolution, 1952–1958." *Specialia* 3 (August 1971): 28–35.

Martínez Alier, Juan and Verena. *Cuba: economía y sociedad*. Paris: Ruedo Ibérico, 1972.

Marx, Karl. *A Contribution to the Critique of Political Economy*. New York: International Publishers, 1981.

———. *The 18th Brumaire of Louis Bonaparte*. New York: International Publishers, 1981.

———. *The Class Struggles in France, 1848–1850*. In Karl Marx and Frederick Engels, *Collected Works*, 10:45–145. New York: International Publishers, 1978.

———. *The Civil War in France: The Paris Commune*. New York: International Publishers, 1968.

Memoria: II Congreso Nacional de la Federación de Mujeres Cubanas. Havana: Editorial Orbe, 1975.

Memorias: Congreso de la CTC. Havana: Central de Trabajadores de Cuba, 1974.
Memorias XIV Congreso de la CTC. Havana: Editorial Orbe, 1980.
Mendoza, Luis G. *Revista Semanal Azucarera: Selecciones, 1935–1945.* Havana: Editorial Lex. 1945.
Mesa-Lago, Carmelo. "The Cuban Economy in the 1980s: The Return of Ideology." In *Socialist Cuba: Past Interpretations and Future Challenges*, edited by Sergio Roca 59–100. Boulder: Westview Press, 1988.
———. *The Economy of Socialist Cuba: A Two Decade Appraisal.* Albuquerque: University of New Mexico Press, 1981.
———. *Cuba in the 1970s: Pragmatism and Institutionalization.* Albuquerque: University of New Mexico Press, 1974.
———. *The Labor Force, Employment, Unemployment and Underemployment in Cuba: 1899–1970.* Beverly Hills: Sage, 1972.
———. *The Labor Sector and Socialist Distribution.* New York: Praeger, 1968.
Mesa-Lago, Carmelo, and Jorge F. Pérez-Lopez. "The Endless Cuban Economic Saga: A Terminal Rebuttal." *Comparative Economic Studies* 27 (Winter 1985): 67–82.
———. "Imbroglios on the Cuban Economy: A Reply to Brundenius and Zimbalist." *Comparative Economic Studies* 27 (Spring 1985): 47–83.
Mills, C. Wright. *Listen Yankee!* New York: Ballantine Books, 1960.
Mintz, Sidney W. "The Rural Proletariat and the Problem of Rural Proletarian Consciousness." *Journal of Peasant Studies* 1 (April 1974): 291–325.
Monreal, Pedro. "Cuba y la nueva economía mundial: el reto de la inserción en América Latina y el Caribe." *Cuadernos de Nuestra América* 8 (January–June 1991): 36–68.
Mora, Alberto. "On Certain Problems of Building Socialism." In *Man and Socialism in Cuba: The Great Debate*, edited by Bertram Silverman, 319–336. New York: Atheneum Press, 1971.
Moreno Fraginals, Manuel. *El ingenio.* 3 vols. Havana: Editorial de Ciencias Sociales, 1978.
Munro, Dana G. *The United States and the Caribbean Republics, 1921–1933.* Princeton: Princeton University Press, 1974.
Nazzari, Muriel. "The 'Woman Question' in Cuba: An Analysis of Material Constraints on Its Solution." *Signs* 2 (1983): 246–263.
Novás Calvo, Lino. "La tragedia de la clase media cubana," *Bohemia Libre* 13 (January 1, 1961): 28–29, 76–77.
Noyola, Juan F. *La economía cubana en los primeros años de la revolución y otros ensayos.* Mexico: Siglo XXI Editores, 1978.
O'Connor, James. *The Origins of Socialism in Cuba.* Ithaca: Cornell University Press, 1970.
Olesen, Virginia. "Context and Posture: Notes on Socio-Cultural Aspects of Women's Roles and Family Policy in Contemporary Cuba." *Journal of Marriage and the Family* 3 (August 1971): 548–560.
Oppenheimer, Andrés. *Castro's Final Hour: The Secret Story Behind the Coming Downfall of Communist Cuba.* New York: Simon & Schuster, 1992.
Ortiz, Fernando. *Contrapunteo cubano del tabaco y el azúcar.* Las Villas: Universidad Central, 1963.
Padula, Alfred L., Jr. "The Fall of the Bourgeoisie: Cuba, 1959–1961." Ph.D. diss., University of New Mexico, 1974.

Page, Charles A. "The Development of Organized Labor in Cuba." Ph.D. diss., University of California, 1952.

Pazos, Felipe. "Comentarios a dos artículos sobre la revolución cubana." *El Trimestre Económico* 29 (January–March, 1962): 1–18.

———. *Influencia de la escuela de ciencias económicas en el desarrollo económico del país*. Santiago de Cuba: Universidad de Oriente, 1955.

———. "Lineamientos de una política de desarrollo económico." *Selva Habanera* 579 (August 13, 1955): 1, 6–7, 9–11.

Pérez, Humberto. "Discurso pronunciado por Humberto Pérez en el acto de clausura del Congreso Constituyente de la Asociación Nacional de Economistas de Cuba." In *Memorias: Congreso Constituyente ANEC*, 119–146. Havana, 1979.

Pérez, Lisandro. "The Family in Cuba." In *The Family in Latin America*, edited by Man Sing Das and Clinton J. Jesser, 235–269. Bombay: Vikas, 1980.

Pérez, Louis A., Jr. *Cuba: Between Reform and Revolution*. New York: Oxford University Press, 1988.

———. *Cuba Under the Platt Amendment, 1902–1934*. Pittsburgh: University of Pittsburgh Press, 1986.

———. "Aspects of Hegemony: Labor, State, and Capital in Plattist Cuba." *Cuban Studies/Estudios Cubanos* 16 (1986): 49–69.

———. *Cuba Between Empires, 1878–1902*. Pittsburgh: University of Pittsburgh Press, 1983.

———. *Army Politics in Cuba, 1898–1958*. Pittsburgh: University of Pittsburgh Press, 1976.

Pérez de la Riva, Juan. "Cuba y la migración antillana." In *Anuario de estudios cubanos: la república neocolonial*, 2:5–75. Havana: Editorial de Ciencias Sociales, 1979.

———. "Los recursos humanos de Cuba al comenzar el siglo: inmigración, economía y nacionalidad (1899–1906)." In *Anuario de estudios cubanos: la república neocolonial*, 2:11–44. Havana: Editorial de Ciencias Sociales, 1975.

Pérez-Lopez, Jorge F. *Measuring Cuban Economic Performance*. Austin: University of Texas Press, 1987.

———. "An Index of Cuban Industrial Output, 1930–1958." In *Quantitative Latin American Studies: Methods and Findings*, edited by James W. Wilkie and Kenneth Riddle, 37–72. Los Angeles: University of California Latin American Center Publications, 1977.

Pérez-López, Jorge F., and Carmelo Mesa-Lago. "Cuba: Counter Reform Accelerates Crisis." *ASCE Newsletter* (March 1991): 3–5.

Pérez-Stable, Marifeli, "The Field of Cuban Studies." *Latin American Research Review* 1 (1991): 239–250.

Pino-Santos, Oscar. *El asalto a Cuba por la oliqarguía financiera yanqui*. Havana: Casa de las Americas, 1973.

Plan de acción contra las irregularidades administrativas y los errores y debilidades del Sistema de Dirección de la Economía. July 17, 1986.

Plan Trienal de Cuba. Havana: Cultural, 1938.

Plenaria Nacional de Chequeo sobre el Sistema de Dirección y Planificación de la Economía. Havana: JUCEPLAN, 1979.

. . . *Porque en Cuba sólo ha habido una revolución.* Havana: Departamento de Orientación Revolucionaria, 1975.

Portes, Alejandro, and John Walton. *Labor, Class and the International System.* Orlando: Academic Press, 1981.

Poyo, Gerald E. *"With All, and for the Good of All": The Emergence of Popular Nationalism in the Cuban Communities of the United States, 1848–1898.* Durham: Duke University Press, 1989.

Primer Congreso del Partido Comunista de Cuba. *Tésis y resoluciones.* Havana: Departamento de Orientación Revolucionaria, 1976.

Proyecto de tésis V congreso de la Federación de Mujeres Cubanas.

Quesada, Felino. "La autonomía de la empresa en Cuba y la implantación del Sistema de Dirección de la Economía." *Cuestiones de la Economía Planificada* 1 (1980): 91–99.

Rabkin, Rhoda P. *Cuban Politics: The Revolutionary Experiment.* New York: Praeger, 1991.

Radaelli, Ana María. "For the Full Equality of Women." *Cuba International* (July 1985): 13–17.

Raggi Ageo, Carlos Manuel. "Contribución al estudio de las clases medias en Cuba." In *Materiales para el estudio de la clase media en la América Latina,* edited by Theo R. Crevenna, 2:73–89. Washington, DC: Panamerican Union, 1950–1951.

Rathbone, John Paul. "Cuba: Current Economic Situation and Short-Term Prospects." *La Sociedad Económica Bulletin* 20 (1992): 1–8.

Ritter, Archibald R. M. "The Organs of People's Power and the Communist Party: The Nature of Cuban Democracy." In *Cuba: Twenty-Five Years of Revolution, 1959–1984,* edited by Sandor Halebsky and John M. Kirk, 270–290. New York: Praeger, 1985.

———. *The Economic Development of Revolutionary Cuba: Strategy and Economic Performance.* New York: Praeger, 1974.

Robinson, Joan. "Cuba: 1965." *Monthly Review* 18 (February 1966): 10–18.

Roca, Blas. *29 artículos sobre la revolución cubana.* Havana: Tipografía Ideas, 1960.

Roca, Blas, and Lázaro Peña. *La colaboración entre obreros y patronos.* Havana: Ediciones Sociales, 1945.

Roca, Sergio. "Housing in Socialist Cuba." In *Housing: Planning, Financing, Construction,* edited by Oktay Ural, 62–74. New York: Pergamon Press, 1979.

———. *Cuban Economic Policy and Ideology: The Ten Million Ton Sugar Harvest.* Beverly Hills: Sage, 1976.

Rodríguez, Carlos Rafael. "Sobre la contribución del Ché al desarrollo de la economía cubana." *Cuba Socialista* 33 (May–June 1988): 1–29.

———. *Cuba en el tránsito al socialismo.* Havana: Editora Política, 1979.

Rodríguez, José Luis. "The Cuban Economy Today." *Cuba Business* 3 (April 1989): 11–12.

Ruiz, Ramón Eduardo. *Cuba: The Making of a Revolution.* Amherst: University of Massachusetts Press, 1968.

Sánchez Rebolledo, Adolfo, ed. *La Revolución Cubana.* Mexico: Ediciones Era, 1972.

Santana, Sara M. "The Cuban Health Care System: Responsiveness to Changing

Needs and Demands." In *Cuba's Socialist Economy Toward the 1990s*, edited by Andrew Zimbalist, 115–127. Boulder: Lynne Rienner, 1987.

Schroeder, Susan. *Cuba: A Handbook of Historical Statistics*. Boston: G. K. Hall, 1982.

Second Congress of the Communist Party of Cuba: Documents and Speeches. Havana: Political Publishers, 1981.

Seers, Dudley. *Cuba: The Economic and Social Revolution*. Westport, CT: Greenwood Press, 1975.

Segunda Plenaria Nacional de Chequeo de la Implantación del SDPE. Havana: JUCEPLAN, 1980.

Sheahan, John. *Patterns of Development in Latin America: Poverty, Repression, and Economic Strategy*. Princeton: Princeton University Press, 1987.

Silva León, Arnaldo. *Cuba y el mercado internacional azucarero*. Havana: Editorial de Ciencias Sociales, 1975.

Silverman, Bertram. "The Great Debate in Retrospective: Economic Rationality and the Ethics of Revolution." In *Man and Socialism in Cuba: The Great Debate*, edited by Bertram Silverman, 3–28. New York: Atheneum Press, 1971.

Skocpol, Theda. *States and Social Revolutions: A Comparative Analysis of France, Russia, and China*. Cambridge: Cambridge University Press, 1979.

Smith, Wayne S. *The Closest of Enemies*. New York: Norton, 1987.

Soto, Lionel. *La revolución del 33*. 3 vols. Havana: Editorial de Ciencias Sociales, 1977.

Spalding, Hobart A., Jr. *Organized Labor in Latin America: Historical Case Studies of Workers in Dependent Societies*. New York: New York University Press, 1977.

Stinchcombe, Arthur L. *Theoretical Methods in Social History*. New York: Academic Press, 1978.

Stokes, William S. "The 'Cuban Revolution' and the Presidential Elections of 1948," *Hispanic American Historical Review* 31 (February 1951): 37–79.

Stoner, K. Lynn. *From the House to the Streets: The Cuban Women's Movement for Legal Reform, 1898–1940*. Durham: Duke University Press, 1991.

Stubbs, Jean. *Cuba: The Test of Time*. London: Latin American Bureau, 1989.

Suárez, Andrés. *Cuba: Castroism and Communism, 1959–1966*. Cambridge: MIT Press, 1967.

Sweezy, Paul, and Charles Bettelheim. *On the Transition to Socialism*. New York: Monthly Review Press, 1971.

Swerling, Boris C., "Domestic Control of an Export Industry: Cuban Sugar." *Journal of Farm Economics* 3 (August 1951): 346–356.

Tabares del Real, José A. *Guiteras*. Havana: Editorial de Ciencias Sociales, 1973.

Tellería, Evelio. *Los congresos obreros en Cuba*. Havana: Instituto Cubano del Libro, 1973.

Tercera Plenaria Nacional de Chequeo de la Implantación del SDPE. Havana: JUCEPLAN, 1982.

Thomas, Hugh. *Cuba: The Pursuit of Freedom*. New York: Harper & Row, 1971.

Thomson, Charles A. "The Cuban Revolution: Reform and Reaction." *Foreign Policy Reports* 11 (January 1, 1936): 262–276.

———. "The Cuban Revolution: Fall of Machado." *Foreign Policy Reports* 11 (December 18, 1935): 250–260.

Tilly, Charles. "Does Modernization Breed Revolution?" In *Revolutions: Theoretical, Comparative, and Historical Studies,* edited by Jack A. Goldstone, 47–57. Orlando: Harcourt Brace Jovanovich, 1986.

———. *Big Structures, Large Processes, Huge Comparisons.* New York: Russell Sage Foundation, 1984.

———. *As Sociology Meets History.* Orlando: Academic Press, 1981.

———. *From Mobilization to Revolution.* Reading, MA: Addison-Wesley, 1978.

Torras, Jacinto. *Obras escogidas: 1945–1958.* Havana: Editora Política, 1985.

———. *Obras escogidas: 1939–1945.* Havana: Editora Política, 1984.

Trotsky, Leon. *History of the Russian Revolution.* 3 vols. London: Sphere Books, 1967.

Valdés, Nelson P. "Cuba: Social Rights and Basic Needs." Paper presented to the Inter-American Commission on Human Rights, Washington, DC, February 25, 1983.

Veiga, Roberto. "Clausura del octavo curso para directores de empresas, celebrado en la Escuela nacional de Dirección de la Economía." *Cuestiones de la Economía Planificada* 2 (1980): 24–34.

Wallerstein, Immanuel. *The Modern World System.* Vol. 1. New York: Academic Press, 1974.

Welch, Richard E., Jr. *Response to Revolution: The United States and the Cuban Revolution.* Chapel Hill: University of North Carolina Press, 1985.

White, Byron. *Azúcar amargo: un estudio de la economía cubana.* Havana: Publicaciones Cultural, 1954.

Wickham-Crowley, Timothy P. *Guerrillas and Revolution in Latin America: A Comparative Study of Insurgents and Regimes Since 1956.* Princeton: Princeton University Press, 1992.

Winocour, Marcos. *Las clases olvidadas en la revolución cubana.* Barcelona: Editorial Crítica, 1979.

Wolf, Eric. *Peasant Wars of the Twentieth Century.* New York: Harper & Row, 1969.

Woodward, Ralph Lee, Jr. "Urban Labor and Communism: Cuba." *Caribbean Studies* 3 (October 1963): 17–50.

XV Congreso de la CTC Memorias. Havana: Editorial de Ciencias Sociales, 1984.

Zanetti, Oscar. "El comercio exterior de la república neocolonial." In *Anuario de estudios cubanos: la republica neocolonial,* 47–126. Havana: Editorial de Ciencias Sociales, 1975.

Zeitlin, Maurice. *The Civil Wars in Chile.* Princeton: Princeton University Press, 1983.

———. *Revolutionary Politics and the Cuban Working Class.* New York: Harper & Row, 1970.

Zeitlin, Maurice, and Richard Earl Ratcliff. *Landlords and Capitalists: The Dominant Class of Chile.* Princeton: Princeton University Press, 1988.

Zeitlin, Maurice, and Robert Scheer. *Cuba: Tragedy in Our Hemisphere.* New York: Grove Press. 1963.

Zimbalist, Andrew. "Incentives and Planning in Cuba." *Latin American Research Review* 24 (1989): 65–93.

———. "Cuban Economic Planning: Organization and Performance." In *Cuba: Twenty-Five Years of Revolution, 1959–1984,* edited by Sandor Halebsky and John M. Kirk, 213–230. New York: Praeger, 1985.

————. "On the Role of Management in Socialist Development." *World Development* 9–10 (1981): 971–977.

Zimbalist, Andrew, and Claes Brundenius. *The Cuban Economy: Measurement and Analysis of Socialist Performance.* Baltimore: Johns Hopkins University Press, 1989.

————. "Growth with Equity: Cuban Development in Comparative Perspective." Unpublished paper.

Zimbalist, Andrew, and Susan Eckstein, "Patterns of Cuban Development: The First Twenty-Five Years." In *Cuba's Socialist Economy Toward the 1990s,* edited by Andrew Zimbalist, 7–24. Boulder: Lynne Rienner 1987.

Zuaznábar, Ismael. *La economía cubana en la década del 50.* Havana: Editorial de Ciencias Sociales, 1986.

OFFICIAL PUBLICATIONS

Cuba

Banco Nacional de Cuba. *Memoria: 1958–1959.* Havana: Editorial Lex. 1960.

————. *Memoria: 1949–1950.* Havana: Editorial Lex. 1951.

Comité Estatal de Estadísticas. *Anuario estadístico de Cuba.* 1982–1988.

Informe de la Comisión Coordinadora de la Investigación del Empleo, Sub-Empleo y Desempleo. *Resultados de la Encuesta sobre Empleo, Sub-Empleo y Desempleo en Cuba (mayo 1956-abril 1957).* Havana. January 1958.

Ministerio de Agricultura. *Memoria del Censo Agrícola Nacional.* Havana: P. Fernández, 1951.

República de Cuba. *Censo de población y viviendas 1981.* Havana: Comité Estatal de Estadísticas, 1983.

————. *Censo de población y viviendas 1970.* Havana: Editorial Orbe, 1975.

————. *Censo de población, viviendas y electoral: informe general* (enero 28 de 1953). Havana: P. Fernández, 1955.

————. *Informe general del censo de 1943.* Havana: P. Fernández, 1945.

United States

Foreign Broadcast Information Service. *Latin America 1992.* March 11, September 14, 1992.

————. *Latin America—Cuba: Fourth Congress of the Cuban Communist Party.* October 15, 1991.

Office of Strategic Services. *The Political Significance and Influence of the Labor Movement in Latin America. A Preliminary Survey. Cuba.* Washington, DC. 1945.

U.S. Department of Commerce. *Investment in Cuba.* Washington, DC. 1955.

U.S. Department of Labor. *Foreign Labor Information: Labor in Cuba.* Washington, DC. 1957.

U.S. State Department, *Foreign Relations of the United States, Cuba 1958–1960* Vol. 6. Washington, DC: Government Printing Office, 1991.

NEWSPAPERS, NEWSLETTERS, AND MAGAZINES

Cuba

Boletín of the Asociación Nacional de Industriales de Cuba
Bohemia
Diario de la Marina
Granma
Granma Weekly Review
Juventud Rebelde
El Mundo
Revolución
Trabajadores

United States

Cuba en el mes
CubaINFO Newsletter
Miami Herald
New York Times
El Nuevo Herald

Index